SCIENCE OF PERCUSSION INSTRUMENTS

SERIES IN POPULAR SCIENCE

Editor-in-Chief: Richard J. Weiss

Published

Vol. 1   A Brief History of Light and Those That Lit the Way
by Richard J. Weiss

Vol. 2   The Discovery of Anti-matter: The Autobiography of Carl David Anderson, the Youngest Man to Win the Nobel Prize
by C. D. Anderson

Series in Popular Science – Vol. 3

SCIENCE OF PERCUSSION INSTRUMENTS

THOMAS D. ROSSING
Northern Illinois University

World Scientific
Singapore • New Jersey • London • Hong Kong

Published by
World Scientific Publishing Co. Pte. Ltd.
P O Box 128, Farrer Road, Singapore 912805
USA office: Suite 1B, 1060 Main Street, River Edge, NJ 07661
UK office: 57 Shelton Street, Covent Garden, London WC2H 9HE

British Library Cataloguing-in-Publication Data
A catalogue record for this book is available from the British Library.

SCIENCE OF PERCUSSION INSTRUMENTS
Series in Popular Science — Volume 3

Copyright © 2000 by World Scientific Publishing Co. Pte. Ltd.

All rights reserved. This book, or parts thereof, may not be reproduced in any form or by any means, electronic or mechanical, including photocopying, recording or any information storage and retrieval system now known or to be invented, without written permission from the Publisher.

For photocopying of material in this volume, please pay a copying fee through the Copyright Clearance Center, Inc., 222 Rosewood Drive, Danvers, MA 01923, USA. In this case permission to photocopy is not required from the publisher.

ISBN 981-02-4158-5
ISBN 981-02-4159-3 (pbk)

This book is printed on acid-free paper.

Printed in Singapore by Uto-Print

Foreword

Percussion instruments are amongst the oldest instruments in the world. They are also, undoubtedly, the most universal. The New Grove Dictionary of Musical Instruments has 1,523 different entries for drums alone. Yet, ironically, percussion has been slow to develop and be utilized in Western art music; hence, the lack of Western scientific interest in these instruments. While extensive research and discovery has been done for the voice, strings and wind instruments, very little attention has been paid to percussion instruments. In the music of the 20th century, however, everything has changed. Percussion has become a dominant player in both contemporary classical music, as well as pop and world music.

Percussion is incredibly versatile with the ability to produce a full range of timbre from noise to nearly pure sound and with a dynamic range that is probably only surpassed by electronic music. Percussion is also constantly evolving. And while the standard instruments will be here for a long time to come, newer instruments are being built and applied in various musical settings. The steel pan of Trinidad and Tobago is an example of a highly evolved instrument that was only conceived of in the middle of the 20th century but now holds an important place in musics of the world. The tuning refinement that has been done on steel pans is phenomenal and on par with many classical instruments which are many times older.

Fortunately, Thomas Rossing has chosen to devote a great deal of his research time exploring the vast array of percussion instruments and discovering the multitude of modes of vibration they produce. I had the privilege of working with Dr. Rossing at Northern Illinois University in the 1970s while teaching and pursuing my Master's degree there. I was able to apply the information and techniques that I learned from him in building and tuning my own percussion instruments. This eventually led to the creation of my own company, Woodstock Percussion, which manufactures Woodstock Chimes and other percussion instruments and distributes them throughout the world.

While the information found in Dr. Rossing's *Science of Percussion Instruments* is certainly important to anyone studying the science of sound, it is absolutely invaluable to the performer and instrument builder. Many percussionists intuitively understand the various instruments which they are called upon to play and use that knowledge to produce many different sounds on each and every one. So, in some cases, Dr. Rossing is merely confirming what percussionists already know intuitively. However, between the covers of his book are a number of wonderful surprises which undoubtedly will affect the way we approach our instruments. Science of Percussion Instruments can be used as a reference tool or read in its entirety. How you apply the wealth of information it contains is up to you. Enjoy, and best of luck in your journey.

Garry Kvistad
CEO, Woodstock Percussion, Inc.
Shokan, New York
January 2000

Preface

Although percussion instruments may be oldest musical instruments (with the exception of the human voice), relatively little has been written about scientific research on these instruments. By way of contrast, string and wind instruments have been the subjects of several good scientific books in recent years. Because the sounds of percussion instruments change so rapidly with time, their study and analysis require equipment that wasn't widely available until quite recently.

I began to study the science of percussion instruments some 25 years ago when Garry Kvistad, who was teaching percussion at Northern Illinois University at that time, asked me some interesting questions about their behavior. Garry and I did some experiments together, and we had many interesting discussions, some of which he carried with him, I believe, when he started his own company. Meanwhile we have studied the acoustics of a wide variety of percussion instruments. Many of them are discussed in *The Physics of Musical Instruments* (Springer-Verlag, New York, 1991, 1998).

Studying the science of percussion instruments has taken me all over the world and has put me in touch with a large number of interesting people: performers, teachers, instrument builders, and other scientists. Besides Garry Kvistad, I would especially like to mention Jacob Malta and André Lehr. Jake Malta, founder of Malmark, Inc., has been a friend for many years. André Lehr, who I consider to be the world's foremost authority on bells, has retired from the Royal Eijsbouts foundry but still devotes much time and effort to the National Carillon Museum in Asten in The Netherlands.

Is it necessary for a musician or a musical instrument builder to understand the science of their instrument? I would argue that it is if they want to compete with the best of their trade. Most builders of fine instruments have mastered the science of their instruments, although in many cases they have done it rather painfully by trial and error. Likewise, skilled performers have learned the science of their instruments by experience. I often remind my students that Stradivari knew all about the physics of violins but it took 300 years to learn it! It is my hope that studying this book will shorten the learning curve for both instrument builders and performers.

This book is written primarily for musicians, but it should be of interest to students of science as well. I have kept the mathematics as simple as possible by translating ideas from the language of physics (mathematics) to non-mathematical language. Readers who wish to go beyond the simple ideas in this book can easily follow the references to more scientific books and to the original scientific literature. Where some principles of physics or perception are necessary to understanding the concepts, these principles are briefly presented in "interludes."

Needless to say, I welcome comments from readers. Who knows, some of these comments may lead to further research. Happy reading.

Thomas D. Rossing
DeKalb, Illinois, 1999

Contents

Chapter 1. The Percussion Family   1
    1.1. The Percussion Family
    1.2. Historical Notes
    1.3. Percussion Ensembles
    References

Chapter 2. Drums with Definite Pitch   5
    2.1. Vibrations of Strings: A Little Bit of Physics
    2.2. Vibrations of Membranes: Key to Understanding Drums
    2.3. Timpani
    2.4. Timpani Sound
    2.5. Interlude: Subjective Tones and Pitch of the Missing Fundamental
    2.6. The Kettle
    2.7. Interlude: Sound Radiation
    2.8. Tabla and Mrdanga
    2.9. Acoustics of Indian Drums
    References

Chapter 3. Interlude: Sound and Hearing   21
    3.1. Sound Waves
    3.2. Hearing Sound
    3.3. Loudness and Musical Dynamics
    3.4. Sound Power Level
    3.5. Masking Sounds
    3.6. Loudness and Duration
    References

Chapter 4. Drums with Indefinite Pitch   26
    4.1. Tom Toms
    4.2. Interlude: Pitch Glide in Membranes
    4.3. Interlude: Modes of a Two-Mass Vibrator
    4.4. Snare Drum
    4.5. Bass Drum
    4.6. Conga Drums
    4.7. Bongos and Timbales
    4.8. Rototoms
    4.9. Irish Bodhrán
    4.10. African Drums
    4.11. Japanese Drums
    4.12. Indonesian Drums
    References

Chapter 5. Interlude: Vibrations of Bars and Air Columns   47
    5.1. Transverse Vibrations of a Bar or Rod
    5.2. Longitudinal Vibrations of a Bar or Rod
    5.3. Torsional Vibrations of a Bar or Rod
    5.4. Vibrations of Air Columns
    5.5. End Correction
    References

Chapter 6. Xylophones and Marimbas   52
    6.1. Xylophones
    6.2. Marimbas
    6.3. Tuning the Bars
    6.4. Resonators
    6.5. Marimba Orchestras and Clair Musser
    6.6. Mallets
    References

Chapter 7. Metallophones   64
    7.1. Orchestra Bells or Glockenspiel
    7.2. Celesta
    7.3. Vibraphone or Vibes
    7.4. Interlude: Thick Bars vs Thin Bars
    7.5. Chimes or Tubular Bells
    7.6. Triangles and Pentangles
    7.7. Gamelan Metallophones
    7.8. Wind Chimes
    7.9. Tubaphones, Gamelan Chimes, and Other Tubular Metallophones
    7.10. African Lamellaphones: Mbira, Kalimba, Likembe, Sanza, Setinkane
    References

Chapter 8. Interlude: Vibrations of Plates and Shells   79
    8.1. Waves in a Thin Plate
    8.2. Circular Plates
    8.3. Elliptical Plates
    8.4. Rectangular Plates
    8.5. Cylindrical Shells
    8.6. Shallow Spherical Shells
    8.7. Nonlinear Effects in Plates and Shells
    References

Chapter 9. Cymbals, Gongs, and Plates   89
    9.1. Cymbals
    9.2. Vibrational Modes in Cymbals
    9.3. Cymbal Sound
    9.4. Nonlinear Behavior of Cymbals

Contents   xi

    9.5. Tam-Tams
    9.6. Gongs
    9.7. Chinese Opera Gongs
    9.8. Bronze Drums
    9.9. Crotales
    9.10. Kyezee
    9.11. Bell Plates
    9.12. Musical Saw
    9.13. Flexatone
    References

Chapter 10. Music from Oil Drums: Caribbean Steelpans   **107**
    10.1. Construction and Tuning
    10.2. Normal Modes of Vibration
    10.3. Interlude: Holographic Interferometry
    10.4. Modes of a Tenor Pan
    10.5. Modes of a Double-Second Pan
    10.6. Sound Spectra
    10.7. Note Shapes
    10.8. Metallurgy and Heat Treatment
    10.9. Skirts
    10.10. Pans of Other Sizes
    10.11. Recent and Future Developments
    References

Chapter 11. Church Bells and Carillon Bells   **128**
    11.1. The Carillon
    11.2. Vibrational Modes of Church Bells and Carillon Bells
    11.3. Tuning and Temperament
    11.4. The Strike Note
    11.5. Major-Third Bells
    11.6. Scaling of Bells
    11.7. Sound Decay and Warble
    11.8. Sound Radiation
    11.9. Clappers
    11.10. Bell Metal
    References

Chapter 12. Handbells, Choirchimes, Crotals, and Cow Bells   **146**
    12.1. Vibrational Modes of Handbells
    12.2. Sound Radiation
    12.3. Sound Decay and Warble in Handbells
    12.4. Timbre and Tuning of Handbells
    12.5. Scaling of Handbells
    12.6. Bass Handbells
    12.7. Choirchimes

12.8. Chinese Qing
12.9. Crotals
12.10. Cow Bells
References

Chapter 13. Eastern Bells — 164
13.1. Ancient Chinese Two-Tone Bells
13.2. Vibrational Modes of Ancient Two-Tone Bells
13.3. Intervals Between the Two Tones
13.4. Temple Bells in China
13.5. Korean Bells
13.6. Japanese Bells
13.7. Other Asian Bells
References

Chapter 14. Glass Musical Instruments — 182
14.1. The Glass Harmonica
14.2. Vibrational Modes of a Wineglass
14.3. Rubbing, Bowing, Striking
14.4. Selecting and Tuning the Glasses
14.5. Verrophone
14.6. Glass Bells
14.7. The Glass Orchestra
14.8. Glass Instruments of Harry Partch and Jean-Claude Chapuis
14.9. Other Glass Instruments
References

Chapter 15. Other Percussion Instruments — 192
15.1. Anklung
15.2. Deagan Organ Chimes
15.3. Other Deagan Instruments
15.4. Instruments of Harry Partch
15.5. Mark Tree
15.6. Instruments of Bernard and François Baschet
15.7. Lithophones
15.8. Ceramic Instruments of Ward Hartenstein
15.9. Thunder Sheet
15.10. Typewriter
References

Name Index — 203

Subject Index — 206

Chapter 1
The Percussion Family

Percussion instruments may be our oldest musical instruments (with the exception of the human voice), but recently they have experienced a new surge in interest and popularity. Many novel percussion instruments have been developed recently and more are in the experimental stage. What is often termed "contemporary sound" makes extensive use of percussion instruments. Yet, relatively little material has been published on the acoustics of percussion instruments.

So reads the introduction to the chapter on percussion instruments in a textbook *The Science of Sound*, the first edition of which I wrote some twenty years ago. In the meantime we have studied the acoustics of many percussion instruments in our laboratory: timpani, snare drums, handbells, gongs, tamtams, cymbals, steelpans, and other instruments. Nevertheless, these words are still true; relatively little material has been published on the acoustics of percussion instruments.

1.1. The Percussion Family

The term percussion means "struck" and strictly speaking percussion instruments are those in which sound is produced by striking. However, the percussion section in a modern orchestra employs many instruments that do not depend upon striking a blow. Indeed, the percussion section is expected to create any unusual sound effect that a composer has in mind. New instruments are constantly being invented and added to the percussionist's repertoire.

There are several ways that have been used to classify percussion instruments. Sometimes they are classified into four groups: *idiophones* (xylophone, marimba, chimes, cymbals, gongs, etc.); membranophones (drums); aerophones (whistles, sirens, etc.); and chordophones (piano, harpsichord). There may be differences of opinion as to whether aerophones and chordophones properly belong in the percussion family. Whistles and sirens are generally played by percussionists in the orchestra; the piano and harpsichord are not. At any rate, this book deals mainly with idiophones and membranophones.

Another way of classifying percussion instruments is by whether or not they convey a definite sense of pitch. Idiophones that convey a definite pitch include bells, chimes, xylophones, marimbas, gongs, and steelpans. Membranophones that convey a definite pitch include timpani, tabla, and mrdanga. Sometimes we described percussive sounds as having a "high" or "low" pitch even if they do not convey an identifiable pitch, but it would be more correct to describe this as high or low range or tessitura.

Percussionists in a modern orchestra or band may have hundreds of instruments to play. Generally the timpanist plays only the timpani, but the other percussionists divide the remaining instruments depending upon the demands of the music. Some works require as many as ten or more percussionists; Schoenberg's *Gurrelieder*, for example, calls for two timpanists and ten other percussionists.

Percussion instruments generally use one or more of the following basic types of vibrators: strings, bars, membranes, plates, air columns, or air chambers. The first four are mechanical; the latter two are pneumatic. Two of them (the string and the air column) tend to produce harmonic overtones; bars, plates, and membranes, in general, do not. The inharmonic overtones of complex vibrators give percussion instruments their distinctive timbres.

This book on the science of percussion instruments considers a large number of instruments: how they vibrate and how they produce sound. In order to understand this, we must consider some basic physics of vibrating systems as well as some psychoacoustics of hearing and perception. This will be done by inserting, when needed, sections or chapters dealing with fundamental principles. Often these are labeled as "interludes." The musician without much previous study of the scientific principles may wish to refer back to these interludes from time to time as the book is being read.

1.2. Historical Notes

Most natural systems follow some type of rhythm: beating hearts, the motion of the planets, ocean waves, phases of the moon, the seasons, the list is long. It is only natural that primitive humans would begin striking sticks or stones together rhythmically. Rhythm is one of the key ingredients in music, and percussion instruments often establish and maintain the rhythm in the performance of music.

One of the best histories of percussion instruments is *Percussion Instruments and Their History* by renowned percussionist James Blades.[1] This book traces percussion instruments from their primitive origins to composers' use of modern percussion.

Blades points out that the earliest instruments were probably idiophones, instruments made of naturally sonorous material which can produce sound without the addition of stretched skin or column of air. These are of five types: shaken idiophones (rattles), stamped idiophones (pits, boards, hollow tubes); scraped idiophones (notched sticks or rasps); concussion idiophones (pairs of similar items such as sticks); and struck idiophones (one or more pieces of sonorous material struck with a stick or bone).

Early in our lives we learned to play with rattles. It is interesting that rattles are among the earliest of percussion instruments. The gourd rattle, a seed pod in which the dried seeds remain, was widely used in primitive societies, especially in Africa. Rattles are still popular in orchestras and ensembles, especially for the performance of Latin American music.

Scraped instruments are found as far back as the early Stone Age. A stick could be drawn across a notched stone, bone, shell, or gourd to produce a raspy sound. The bone scraper has been closely associated with the hunt, erotic rituals, and funeral ceremonies. Scrapers were found among Indian tribes in North and South America, and also in Africa.

The earliest drums were probably log drums of various types. Later, it was discovered that by stretching an animal skin across the cavity in the log, a louder sound could be made. Eventually the membrane drum came to be the most important percussion instrument. The earliest drums were probably struck with the hands, but the use of sticks as beaters was found to increase the loudness of the sound. Later a second membrane was added. Today, there are thousands of different types of drums found throughout the world. Throughout the years, drums have been used for signaling, for sending messages, and for marshaling troops to battle as well as for performing music.

1.3. Percussion Ensembles

Although percussion instruments have most often been used in ensemble with string and wind instruments, a number of successful ensembles have relied on percussion instruments alone. Sometimes these ensembles use one type of instrument, such as steel bands (see Chapter 10) and marimba orchestras (see Chapter 6), more often they employ a variety of percussion instruments, such as the Black Earth ensemble, shown in Fig. 1.1, whose members were once artists-in-residence at our university. A steel band is shown in Fig. 1.2.

Fig. 1.1. The Black Earth Percussion Ensemble at Northern Illinois University used a large variety of percussion instruments, as can be seen in this photo.

Fig. 1.2. The Northern Illinois University Steel Band is an example of a percussion ensemble using one type of percussion instrument.

References

1. J. Blades, *Percussion Instruments and their History* (Faber and Faber, London, 1974).

Chapter 2
Drums with Definite Pitch

Drums have played an important role in nearly all musical cultures. They have been used to transmit messages, convey the time of day, send soldiers into battle, and warn of impending danger. Drums are practically as old as the human race.

The earliest drums were probably chunks of wood or stone placed over holes in the earth. Then it was discovered that more sound could be obtained from hollow tree trunks, the ancestors of our contemporary log drums. The most familiar type of drum consists of a membrane of animal skin or synthetic material stretched over some type of air enclosure.

In this chapter, we will discuss drums that convey a strong sense of pitch, including kettledrums, tabla, roto-toms, and boobams. We will learn that the modes of vibration of these drums have frequencies which are nearly harmonics of a fundamental, and that is why we hear a definite pitch.

2.1. Vibrations of Strings: A Little Bit of Physics

A guitar string is probably the simplest of all musical vibrators. Yet its vibrations can be deceptively complex. When drawn to one side and released, the string vibrates in a rather complex way that can be described as a combination of normal modes of vibration. For example, if it is plucked at its center, the nearly triangular shape it assumes, as it vibrates to and fro, can be thought of as being made up of simple modes having frequencies that correspond to the fundamental plus the odd-numbered harmonics, as listed in Fig. 2.1.

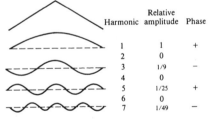

Fig. 2.1. Odd-numbered modes of vibration add up in appropriate amplitude and phase to the shape of a string plucked at its center.

Harmonic	Relative amplitude	Phase
1	1	+
2	0	
3	1/9	−
4	0	
5	1/25	+
6	0	
7	1/49	−

Figure 2.1 illustrates how the modes associated with the odd-numbered harmonics, when each is present in the right proportion, can add up at one instant to give the initial shape of the string. Modes 3, 7, 11, etc. must be opposite in phase from modes 1, 5, and 9 in order to give a maximum in the center. Also, the relative amplitudes are in the ratios $1/n^2$, where n is the number of the mode.

The force necessary to restore the string to its center (equilibrium) position when it is displaced comes from the force or tension applied to the string. To tune the string to a higher frequency, the guitarist increases the tension by means of the tuning machines or pegs. Increasing the tension increases the frequencies of all the various modes of vibration but maintains their harmonic ratios.

2.2. Vibrations of Membranes: Key to Understanding Drums

A membrane can be thought of as a two-dimensional string, in that its restoring force is due to tension applied from the edge. A membrane, like a string, can be tuned by changing the tension. Membranes, being two-dimensional, can vibrate in many modes that are not normally harmonic; that is, the frequencies of the higher modes are not simple integers times the fundamental frequency.

Four modes of a circular membrane are shown in Fig. 2.2. In the first mode (the fundamental), the entire membrane moves in the same direction, although the center moves through the greatest amplitude. In the other modes, there are one of more nodal circles or nodal diameters that act as boundaries or pivot lines. The parts of the membrane on either side of a nodal line move in opposite directions.

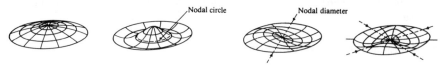

Fig. 2.2. Modes of a circular membrane. The first two modes have circular symmetry; the second two do not.

A membrane, like a string, can be tuned by changing its tension. A major difference between vibrations in a membrane and in a string, however, is that while the mode frequencies in a string are harmonics of the fundamental, in a two-dimensional membrane they are not. Another difference is that in a membrane, nodal lines (circles and diameters) replace nodal points along the string.

Nodal circles and nodal diameters in the first 12 modes of a membrane are shown in Fig. 2.3. A nodal circle always occurs at the edge of the membrane where it is supported. Above each diagram are given two numbers that designate the number of nodal diameters and circles, respectively. For example, the (21) mode has two nodal diameters and one nodal circle (at the edge).

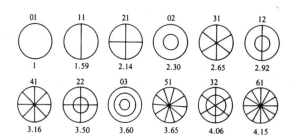

Fig. 2.3. Vibrational modes of a membrane, showing radial and circular nodes and the customary mode designation (the first number gives the number of radial modes, and the second the number of circular nodes, including the one at the edge). The number below each mode diagram gives the frequency of that mode compared to the fundamental (01) mode.

In either a string or a membrane, the modal frequencies vary as the square root of the tension. Thus to double the frequency, the tension would have to be quadrupled, which is quite impractical. A more practical example would be that to raise the frequency by 6 % (corresponding to a semitone on the musical scale) the tension would have to increase by about 12%.

Actually the frequency of a membrane is determined by the ratio of tension to mass per unit area, so increasing the thickness of a drumhead by 12% lowers the mode frequencies by 6% just as increasing the tension by 12% raises it by the same amount. Increasing the radius by 12%, on the other hand, decreases the mode frequencies by a full 12%.

2.3. Timpani

The timpani or kettledrums are the most frequently used drums in the orchestra, one member of the percussion section usually devoting attention exclusively to them. During the last century, various mechanisms were developed for changing the tension to tune the drumheads rapidly. Most modern timpani have a pedal-operated tensioning mechanism in addition to six or eight tensioning screws around the rim of the kettle. The pedal typically allows the player to vary the tension over a range of 2:1, which corresponds to a tuning range of about a musical fourth. A modern pedal-equipped kettledrum is shown in Fig. 2.4.

Fig. 2.4. Kettledrum.

At one time all timpani heads were calfskin, but this material has gradually given way to Mylar (polyethylene terephthalate). Calfskin heads require a great deal of hand labor to prepare and great skill to tune properly. Some orchestral timpanists prefer them for concert work under conditions of controlled humidity, but use Mylar when touring. Mylar is insensitive to humidity and easier to tune, due to its homogeneity. A thickness of 0.19 mm (0.0075 inch) is considered standard for Mylar timpani heads. Timpani kettles are roughly hemispherical; copper is the preferred material, although fiberglass and other materials are also used.

Although the modes of vibration of an ordinary membrane are not harmonic in frequency, a carefully tuned kettledrum sounds a strong fundamental plus two or more harmonic overtones. Lord Rayleigh [1] recognized the principal note as coming from the (11) mode and identified overtones about a perfect fifth (3:2 frequency ratio), a major seventh (15:8 frequency ratio), and an octave (2:1 frequency ratio) above the principal tone. Timpanist H. W. Taylor [2] identified a tenth (octave plus a third, 5:2 frequency ratio) by humming near the drumhead, a technique some timpanists use to fine-tune their instruments.

How are the inharmonic modes of an ordinary membrane coaxed into a harmonic relationship in the timpani? Three effects contribute:
(1) The membrane vibrates in a "sea of air," and the mass of this air lowers the frequency of the vibrational modes, especially those of low frequency;

(2) The air enclosed by the kettle has resonances of its own that interact with the modes of the membrane that have similar shapes;
(3) The stiffness of the membrane raises the frequencies of the higher overtones.
Our studies show that the first effect (air loading) is mainly responsible for establishing the harmonic relationship of kettledrum modes; the other two effects only "fine tune" the frequencies but may have considerable effect on the decay rate of the sound [3].

2.4. Timpani Sound

Vibration frequencies of a kettledrum, a drumhead with the kettle, and an "ideal" (unloaded) membrane are given in Table 2.1 along with ratios to the principal (11) mode. Note that the enclosed air in the kettle raises the frequencies of the (01), (02), and (03) modes which are circularly symmetrical. The (11), (21), (31), and (41) modes in the kettledrum have frequencies in the ratios 1 : 1.5 : 1.99 : 2.44, which is sufficiently close to the harmonic ratios 1 : 1.5 : 2 : 2.5 to give the kettledrum a strong sense of pitch. To preferentially excite these modes, the timpanist generally strikes the head about one-fourth of the way from the edge to the center. Striking the head at its center preferentially excites the rapidly-decaying (01) mode (along with the (02) and (03) modes, of course), producing a "thud."

Table 2.1. Vibration frequencies of a 65-cm (26-inch) kettledrum, a drumhead without the kettle, and an ideal membrane (perfectly flexible and unloaded by air)

Mode	Kettledrum		Drumhead alone		Ideal membrane
	f	f/f_{11}	f	f/f_{11}	f/f_{11}
01	127 Hz	0.85	82 Hz	0.53	0.63
11	150	1.00	155	1.00	1.00
21	227	1.51	229	1.48	1.34
02	252	1.68	241	1.55	1.44
31	298	1.99	297	1.92	1.66
12	314	2.09	323	2.08	1.83
41	366	2.44	366	2.36	1.98
22	401	2.67	402	2.59	2.20
03	418	2.79	407	2.63	2.26
51	434	2.89	431	2.78	2.29
32	448	2.99	479	3.09	2.55
61	462	3.08	484	3.12	2.61
13	478	3.19	497	3.21	2.66
42			515	3.32	2.89

Sound spectra obtained by striking a 65-cm (26-inch) diameter kettledrum in its normal place (about one-fourth of the way from the edge to the center) and at the center are shown in Fig. 2.5. Note that the (0,1) mode appears much stronger when the drum is struck at the center, as do the other symmetrical modes [(0,2) and (0,3)]. These modes die out rather quickly, however, so they do not produce a very clear sound. In fact, striking the drum at the center produces quite a dull, thumping sound.

Drums with Definite Pitch

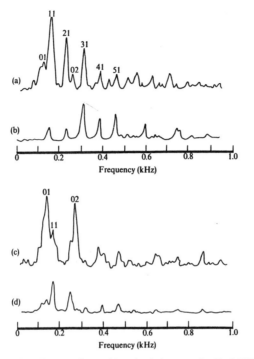

Fig. 2.5. Sound spectra from a 65-cm kettledrum tuned to E_3 (165 Hz):
(a) approximately 0.03 s after striking at the normal point;
(b) approximately 1 s later;
(c) approximately 0.03 after striking at the center;
(d) approximately 1 s later [4].

2.5. Interlude: Subjective Tones and Pitch of the Missing Fundamental

At this point, we would like to take time to describe a very interesting psychoacoustical effect that has important applications in the perception of sound and music. When the ear is presented with a tone composed of exact harmonics, it is easy to predict what pitch will be heard. It is simply the lowest common factor in these frequencies, which is the fundamental. The ear identifies the pitch of the fundamental, even if the fundamental is very weak or missing altogether. For example, if the ear hears a tone having partials with frequencies of 600, 800, 1000, and 1200 Hz, the pitch will nearly always be identified as that of a 200-Hz tone, the "missing fundamental." This is an example of what is called *virtual pitch*, since the pitch doesn't correspond to any partial in the complex tone. The ability of the ear to determine a virtual pitch makes it possible for the undersized loudspeaker of a portable radio to produce bass tones, and it also forms the basis for certain mixture stops on a pipe organ.

If a strong fundamental is not essential for perceiving the pitch of a musical tone, the question arises as to which harmonics are most important. Experiments have shown that for a complex tone with a fundamental frequency up to about 200 Hz the pitch is mainly determined by the fourth and fifth harmonics. As the fundamental frequency increases, the number of the dominant harmonics decreases, reaching the fundamental itself for f_0 =2500 Hz and above (Plomp, 1976). When the partials of the complex tone are not harmonic, however, the determination of virtual pitch is more subtle. The ear can apparently pick out a series of nearly harmonic partials somewhere near the center of the audible range, and determine the pitch to be the largest near-common factor in the series (see Chapter 7 in ref. 6). Several demonstrations of virtual pitch are presented on a Compact Disc by Houtsma, Rossing, and Wagenaars [5].

Musical examples of virtual pitch from "near harmonics" in a complex tone are the sounds of bells and chimes. In each case, the pitch of the strike note is determined mainly by three partials that have frequencies almost in the ratio 2:3:4, as we will discuss in Chapter 11. In the case of the bell, there is usually another partial with a frequency near that of the strike note which reinforces it. In the case of the chime, however, there is none: The pitch is purely subjective.

Does the Timpani have a Virtual Pitch?

Nearly all listeners identify the pitch of the timpani as corresponding to the frequency of the (1,1) partial. It is a little surprising, perhaps, that the pitch is that of the principal partial rather than the missing fundamental of the nearly-harmonic series of partials, which have the ratios 1 : 1.5 : 1.99 : 2.44 for the timpani in Table 2.1. The nearest common factor is clearly 0.5, so the virtual pitch would be an octave below the principal partial, or 75 Hz for the drum in Table 5.1, rather than the 150 Hz that most listeners hear.

Apparently the strengths and durations of the overtones are insufficient, compared to the principal tone, to establish the harmonic series of the missing fundamental. Some timpanists report that a gentle stroke at the proper spot with a soft beater can produce a rather indistinct sound an octave below the nominal pitch [7].

2.6. The Kettle

From Table 2.1, it is clear that the frequencies of the important (11), (21), (31), and (41) modes are not much influenced by the kettle. Without the kettle, these mode frequencies had ratios of 1.00:1.47:1.91:2.36, not quite as harmonic as the corresponding mode frequencies with the kettle but quite tolerable. Even without the kettle, an air-loaded timpani membrane conveys a fairly definite sense of pitch.

The main role of the kettle is that of a baffle to acoustically separate the top and bottom sides of the membrane. This increases the radiation efficiency which decreases the decay times, especially in the modes of lower frequency. This is similar to the acoustical effect of an enclosure on a loudspeaker; without the enclosure radiation efficiency is diminished at the lower frequencies. (A simple demonstration experiment that illustrates this is described in Chapter 20 of ref. 6).

Drums with Definite Pitch

Vibration of Air Enclosed by the Kettle

One expects the air enclosed by the kettle to vibrate in a series of normal modes that can couple to membrane modes having maxima in the same relative positions, just as the enclosed air in a guitar body, for example, couples to the top and back plates [6].

In order to better understand how the kettle fine tunes the vibrational modes of the membrane, we did a series of experiments to determine the vibrational modes of the air enclosed by the kettle. In order to do this, we covered the kettle with a rigid cover having small holes through which a driving tube and probe microphone could be inserted. In the top line of Fig. 2.6 are the normal modes of the air in a rigidly capped 65-cm (26-inch) diameter kettle. In every case, the kettle air modes are higher in frequency than the membrane modes (see Table 2.1) to which they couple.

Also shown in Fig. 2.6 are the corresponding air modes in the kettle when the volume is reduced to one-half and one-quarter of the original volume by partly filling the kettle with water. Note that some air modes are raised in frequency while some are lowered by reducing the kettle volume.

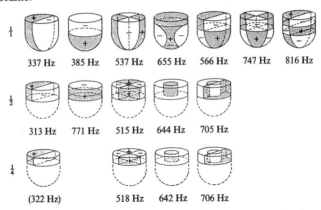

Fig. 2.6. Normal modes of the air enclosed in a rigidly capped kettle for three different air volumes.

Is Kettle Volume Important in Determining Timpani Sound? Yes.

The effect of kettle volume and shape on timpani sound was the subject of a collaborative study carried out at Northern Illinois University and Purdue University. Experimental studies included measurements of sound spectra and decay times with the kettle reduced in volume by adding water, as in the experiments on air modes described in the previous section. The experimental results agreed very well with theoretical calculations made using mathematical Green function [8].

The results of these studies (summarized in Chapter 18 of ref. [9]) showed, not surprisingly, that modern timpani have kettles of just about the right volume to optimize the harmonicity of the principal partials. A lot of trial and error and careful design has gone into modern timpani.

Is Kettle Shape Important in Determining Timpani Sound? No.

Careful studies indicate that the shape of the timpani kettle is quite unimportant in determining timpani sound, provided the volume is kept in the correct range [10]. These results may surprise some timpanists, since articles in the literature frequently refer to "shaping" the sound by using kettles with hemispherical, parabolic, or other shapes. A simple fact that helps explain why kettle shape is unimportant is that the sound wavelengths are so much larger than the kettle dimensions. (At 140 Hz, for example, the wavelength is 2.5 meters, and even at 440 Hz, it is about 0.8 m). Thus the mode frequencies of the enclosed air, which depend on kettle volume, are virtually unaffected by the shape of the kettle.

Another misunderstanding about kettles has to do with the vent hole generally placed at the bottom of the kettle to equalize air pressure inside and outside as the barometric pressure changes (and thus prevent the membrane from bulging up or dishing down). At least one respected source reports that the vent hole has a strong influence on timpani sound through its damping of air vibrations in the kettle [11]. Careful studies show that this is not the case, however [3]. The vent hole is convenient for humidifier pads when using calfskin heads.

Sound Decay With and Without the Kettle

Figure 2.5 shows that decay times are quite long for the (11), (21), (31) and (41) modes but quite short for the (01) and (02) modes. Sound decay times for several modes with and without the kettle are shown in Table 2.2. Following the practice generally employed in architectural acoustics to express reverberation time, τ_{60} is the time for the sound to decrease by 60 dB. Note that times of the (01), (02), and (03) modes are very short with the kettle in place. Decay times for the (11) and (21) modes are shorter with the kettle than without, although an exact comparison is made more difficult by the fact that different membrane tensions led to slightly different mode frequencies in the two experiments.

Table 2.2. Decay times for a 65-cm (26-inch) timpani membrane with and without a kettle

	With kettle				Without kettle			
	$T = 5360\,\text{N/m}$		$T = 3710\,\text{N/m}$		$T = 4415\,\text{N/m}$		$T = 2820\,\text{N/m}$	
Mode m,n	f_{mn} (Hz)	τ_{60} (s)	f_{mn} (Hz)	τ_{60} (s)	f_{mn} (Hz)	τ_{60} (s)	f_{mn} (Hz)	τ_{60} (s)
0,1 monopole	140	< 0.3	128	0.4	93	0.8	73	1.5
1,1 dipole	172	0.8	145	2.3	173	2.5	139	3.4
2,1 quadrupole	258	1.7	218	3.7	249	3.3	204	3.4
0,2	284	0.4	235	0.3	270	0.4	214	< 0.3
3,1	340	2.7	287	4.6	322	2.6	267	4.6
1,2	344	0.5	303	2.5	367	< 0.3	295	1.3
4,1	420	1.7	354	4.3	394	2.8	330	4.2
2,2	493	0.5	394	0.9	445	0.7	364	1.2
0,3	467	< 0.3	383	0.5	458	< 0.3	353	< 0.3
5,1	501	2.6	421	4.1	467	2.1	392	4.2

ª After Christian et al. [7]

Drums with Definite Pitch

On the basis of Table 2.2 and similar data it is possible to make the following observations:
(1) The harmonically tuned (1,1), (2,1), (3,1), and (4,1) modes decay much more slowly than the other modes, especially the (0,1), (0,2), and (0,3) modes.
(2) The (0,1) mode, which acts as a monopole source, decays rapidly; the baffled (1,1) mode and the unbaffled (0,1) mode, which act as dipole sources, decay less rapidly; the baffled (2,1) mode and the unbaffled (1,1) mode decay still less rapidly.

2.7. Interlude: Sound Radiation

The sound field radiated by musical instruments, like other sound sources, can often be approximated by considering radiation from simple sound sources. The simplest source is a monopole source, which radiates equally in all directions. A balloon being inflated and deflated many times a second would be such a source. Almost any source that is much smaller than the wavelength of sound approximates a monopole. The radiation efficiency of a monopole source increases as f^2 which indicates a lesser dependence on frequency than dipoles or higher order sources.

A dipole source could be characterized by two balloons connected by a pipe, so that when one is inflated the other is deflated so that there is no net displacement of air. In the plane that bisects the dipole, the sound field will be zero due to cancellation from the two equally distant balloons. However, at all other locations the radiation from the nearer balloon is greater, and a net sound field results, as shown in Fig. 2.6 b. The radiation efficiency of a dipole source is very small at low frequency, but it increases as f^4 so it becomes more of a factor at higher frequency.

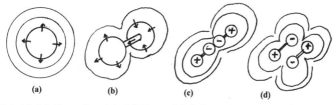

Fig. 2.6. (a) A balloon alternately inflating and deflating acts as a monopole source which radiates sound equally in all directions; (b) Two balloons connected by a pipe, so that one inflates when the other deflates, act as a dipole source which has maximum sound radiation in the axial direction; (c) Two dipoles end to end form a liner quadrupole source; (d) Two parallel dipoles form a tesseral quadrupole source.

Two dipoles can be combined in two different ways to form a quadrupole source, as shown in Fig. 2.6 c,d. A linear quadrupole, shown in Fig. 2.6 c consists of two dipoles end to end, so that the radiated field has a maximum along the axis, as does the field of a dipole, but now the plane of symmetry becomes a weaker maximum. When dipoles are placed side-by-side, as in Fig. 2.6 d, a tesseral quadrupole results, and the sound field has four maxima in the directions of the individual radiators, as shown. For either type of quadrupole, the radiation field is proportional to f^6 so they are effective radiators only at relatively high frequencies.

A timpani membrane vibrating in its lowest 3 modes, with and without a kettle, roughly approximates monopole, dipole, and quadrupole sources, as shown in Fig. 2.7. The (0,1) mode, for example, approximates a monopole source when it vibrates over a kettle (2.7a) but approximates a dipole source (2.7b) when it vibrates without the kettle to act as a baffle. Thus it radiates more efficiently (and loses its vibrational energy more quickly) when the kettle is attached, as we have already discussed.

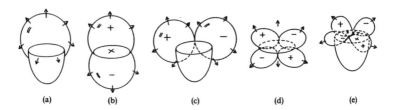

Fig. 2.7. Radiation from a vibrating timpani membrane: (a) (0,1) mode in a membrane baffled by the kettle approximates a monopole; (b) (0,1) mode in an unbaffled membrane (kettle removed) approximates a dipole; (c) (1,1) mode with kettle approximates a dipole; (d) (1,1) mode without baffle (kettle) approximates a quadrupole; (e) (2,1) mode with kettle also approximates a quadrupole.

Graphs showing the directionality of sound radiation from a 40-cm diameter kettledrum are shown in Fig. 2.8. Note that the radiation from the (01) mode is the same in all directions in the plane of the membrane, while the (11) mode radiates more strongly in two directions, the (21) mode in 4 directions, and the (31) mode in 6 directions [12].

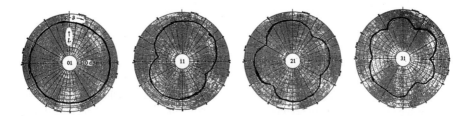

Fig. 2.8. Sound radiation from a 40-cm diameter kettledrum in the plane of the membrane. In these polar graphs, relative sound level L at each angle θ is represented by the distance from the center. The number of maxima is twice the number of nodal diameters.

One lesson to be learned from Fig. 2.8 is that no two persons in the concert hall hear the same musical performance. Musical instruments radiate different sounds in different directions, and so each listener hears a different mix of sounds. Of course sound reflections from walls, ceiling, and other surfaces tend to mix the sounds and reduce the directional dependance. Nevertheless, music critics should realize that the listener a few meters away may be hearing quite a different concert. The directional patterns of musical instruments create quite a problem for recording engineers. These are especially acute if microphones are placed so close to instruments that the direct sound exceeds the reflected sound. Finally, the performer may have the "worst" location in the hall when it comes to hearing the sound from his/her own and nearby instruments.

2.8. Tabla and Mrdanga

Foremost among the drums of India are the *tabla* of North India and the *mrdanga* (also known as mirdangam or mirdang) of South India shown in Fig. 2.9. Both drums convey a strong sense of pitch due to the harmonic tuning of the partials in their sound spectra.

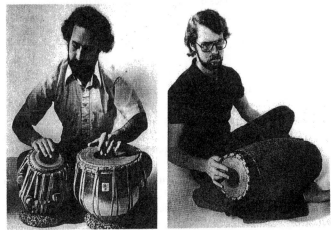

Fig. 2.9. Tabla (left) and mrdanga (right) are traditional drums of North India and South India. Both are tuned by applying starchy paste to the drumheads.

Unlike timpani, which depend upon air loading of the drumhead to shift the inharmonic modes of vibration of an ordinary membrane into a nearly harmonic relationship, tabla and mrdanga load the membrane with a paste of starch, gum, iron oxide, charcoal or other materials. The small diameter of their drumheads make tuning by air loading alone impractical.

The term *tabla* is sometimes used to describe a single drum but more often it describes a pair of drums used in Hindustani (North Indian) classical music. The right-hand drum or *daya* (or dahina, meaning right) is a tunable drum with a range of about an octave.

The body is traditionally carved from a block of wood. The left-hand drum or *bayan* (also bamya, meaning left) is a larger drum, generally with a metal body.

The most unusual feature of the daya is the three-piece design of the head. The main surface of the head (*sur*) is held in place by a collar (*kinar*) of goatskin whose edge is woven into a braid (*gajra*) which is held onto the body with straps. At the center of the drumhead (*pudi*) is a black disk, known as the *syahi* which is carefully shaped to accomplish the harmonic tuning of the partials. The head of the bayan also has a syahi, though it is placed off-center. When played, the palm of the left hand is placed near the center of the bayan, and the player can change the pitch of the drum by changing the pressure on the head.

Tabla players make extensive use of vocal mnemonics for describing compositions. Each stroke is given a one-syllable name. Strokes on the dayan are pronounced toward the front of the mouth (na, ta, te, tin, tun), while strokes on the bayan are pronounced toward the back (ge, ke). This allow fast rolls to be pronounced quickly.

The mrdanga (or mrdangam) is the main drum in the Carnatic tradition in South India. This instrument is a single piece of wood that is hollowed out and has playing heads on both sides so that it functions, in many respects, like the dayon and bayon combined into one. The mridanga uses heavier hides than the tabla, and since they resonate over a common chamber, there is acoustical coupling between the two heads. The larger head of the mridangam is generally loaded with a paste of wheat and water shortly before playing. The tabla often has a string placed between the annular covering and the main skin, while the mridanga uses straw.

2.9. Acoustics of Indian Drums

A succession of Indian scientists, beginning with Nobel laureate C. V. Raman, have studied the acoustical properties of these drums. Raman and his colleagues recognized that the first four overtones of the tabla are harmonics of the fundamental mode. Later they identified these five harmonics as coming from nine normal modes of vibration, several of which have the same frequencies. The fundamental is from the (01) mode; the second harmonic is from the (11) mode; the (21) and (02) modes provide the harmonic; the (31) and (12) modes similarly supply the fourth harmonic; and three modes, the (41), (03), and (2,2) contribute to the fifth harmonic.

Chladni (powder) patterns of six different vibrational modes, all of which have frequencies near the third harmonic, are shown in Fig. 2.10. These patterns, published by Raman in 1934, were skillfully obtained by sprinkling fine sand on the membrane before or immediately after the stroke. The sand gathers along the nodes–the lines of least vibration–forming a map of the vibration pattern excited by that stroke. Fig. 2.10a show the (21) mode and Fig. 2.10f shows the (02) mode; the other four shapes are combinations of these two normal modes obtained by touching the membrane in carefully selected places. Exciting any of the vibrational shapes in Fig. 2.10 will result in a third-harmonic partial.

Drums with Definite Pitch 17

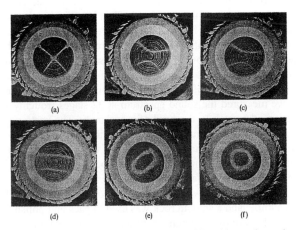

Fig. 2.10. Chladni patterns of six different vibrational shapes of the tabla membrane that all have frequencies near the third harmonic. The (21) and (02) normal modes are shown in (a) and (f), while (b)-(e) are combinations of these two modes [12].

Figure 2.11 show the patterns of the nine normal modes of vibration corresponding to the five harmonics. Also shown are some of the combinations whose vibrational frequencies correspond to the five tuned harmonics.

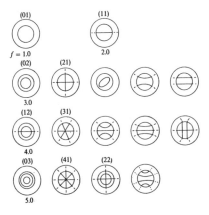

Fig. 2.11. Nodal patterns of the nine normal modes and seven of the combinations that correspond to the five harmonics of the tabla or mrdanga head. Mode designations are given above the patterns and harmonic numbers are given below [13].

Chladni patterns like those in Fig. 2.10 indicate that most of the vibrational energy is confined to the loaded portion of the drumhead. This is accentuated by the restraining action of a rather stiff annular leather flap in loose contact with the peripheral portion of the drumhead, dividing the drumhead into three concentric regions, which can be seen in Fig. 2.10. The player uses these three regions to obtain three distinctly different sounds, described as "tun" (center portion), "tin" (unloaded portion), and "na" (outermost portion).

In order to study the effect of the center patch on the modes of vibration and the sound of a drum, we measured the sound spectrum at 32 stages (roughly each 3 layers) during the application of a patch to the head of a mrdanga. The paste was prepared by kneading together roughly equal volumes of overcooked rice and a black powder composed of manganese and iron oxide. The area at the center of the head was cleaned, dried, and scraped with a knife to raise the nap and provide good adhesion. A thin layer of overcooked rice was first smeared onto this surface as glue, and then a small lump of the paste was applied. It was smeared out evenly with a swirling motion by the thumb. The excess was scraped away with a knife, and a smooth rock was used to pack and polish the mixture by rubbing the surface. The resulting spectral frequencies are shown in Fig. 2.12.

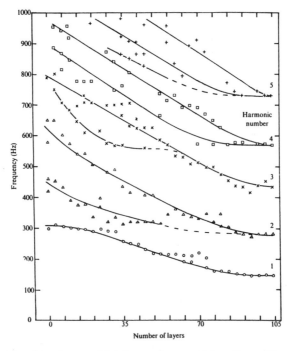

Fig. 2.12. Frequencies of prominent partials in the sound spectra of the mrdanga with no patch and at various stages during application of the patch [13].

Drums with Definite Pitch

Note the five harmonic partials in the sound of the fully loaded membrane in Fig. 2.12. Note also that several of these partials originate from two or three modes of vibration tuned to have the same frequency by appropriately loading the membrane. This is similar to the behavior Raman noted in the tabla.

Studies by Ramakrishna and Sondhi [14] at the Indian Institute of Science in Bangalore and by De [14] in Santiniketan, West Bengal, have indicated that the areal density of the loaded portion of the membrane should be approximately 10 times as great as the unloaded portion (which is typically around 0.02-0.03 g/cm^2). The total mass of the loaded portion is in the range of 9–15 g for different tabla. We estimate that the mrdanga patch in Fig. 2.12, which was about 3 mm thick at the center, had a total mass of 29 g and an area density of 0.8 g/cm^2. The density of the dry paste was about 2.8 g/cm^3.

Sound spectra from the "din" stroke with no patch, with about half the layers in place, and with the finished patch are shown in Fig. 2.13. The differences are rather striking. The fundamental (01) mode has moved down in frequency from 295 Hz to 144 Hz, slightly more than one octave. The other modes have moved by varying amounts into a nearly harmonic frequency relationship.

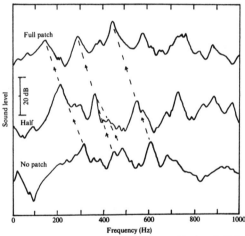

Fig. 2.13. Sound spectra from the *din* stroke with no patch, with about half the layers in place, and with the finished patch [13].

References

1. Lord Rayleigh *The Theory of Sound*, 2nd ed., Vol. I (Macmillan, New York, 1894) p. 348. (Reprinted by Dover, New York, 1945).
2. H. W. Taylor, *The Art and Science of the Timpani* (Baker, London, 1964).
3. T. D. Rossing, *Sci. Am.* **247**(5) (1982) 172.
4. T. D. Rossing and G. Kvistad, *Percussionist* **13** (1976) 90.

5. A. J. M. Houtsma, T. D. Rossing, and W. M. Wagenaars, *Auditory Demonstrations* (Philips Compact Disc #1126-061 and text, 1987).
6. T. D. Rossing, *The Science of Sound,* 2nd ed. (Addison-Wesley, Reading, MA, 1990).
7. R. S. Brindle, *Contemporary Percussion* (Oxford University Press, London, 1970).
8. R. S. Christian, R. E. Davis, A. Tubis, C. A. Anderson, R. I. Mills, and T. D. Rossing, *J. Acoust. Soc. Am.* **76** (1984) 1336.
9. N. H. Fletcher and T. D. Rossing, *The Physics of Musical Instruments* 2nd ed. (Springer-Verlag, New York, 1991).
10. A. H. Benade, *Fundamentals of Musical Acoustics* (Oxford University Press, New York, 1976). Reprinted by Dover,
11. H. Fleischer, *Die Pauke: Mechanischer Schwinger und akustiche Strahler* (Univ. der Bundeswehr, München, 1988).
12. C. V. Raman, *Proc. Indian Acad. Sci.* **A1** (1934) 179. Reprinted in *Musical Acoustics: Selected Reprints*, ed. T. D. Rossing (Am. Assn. Phys. Teach., College Park, MD, 1988).
13. T. D. Rossing and W. A. Sykes, *Percussive Notes* **19**(3) (1982) 58.
14. B. S. Ramakrishna and M. M. Sondhi, *J. Acoust. Soc. Am.* **26** (1954) 523.
15. S. De, *Acustica* **40** (1978) 206.

Chapter 3
Interlude: Sound and Hearing

Since we will be dealing with the sounds of percussion instruments throughout this entire book, it is important to understand what is meant by sound. The word *sound* is actually used to describe two different things: (1) an auditory sensation in the ear; and (2) the disturbance in a medium that can cause this sensation. Making this distinction answers the age-old question, "If a tree falls in a forest and no one is there to hear it, does it make a sound?"

3.1. Sound Waves

Sound is carried through gases, liquids, and solids by sound waves. The world is full of waves. Besides sound waves, there are light waves, water waves, shock waves, radio waves, X-rays, and others. The place you are sitting is being crisscrossed by light waves, radio waves, as well as sound waves of various frequencies. Practically all communication depends on waves of some type. Although sound waves are vastly different from radio waves or ocean waves, all waves possess certain common properties. They are capable of being *reflected, refracted* (bent as they go from one medium to another) and *diffracted* (spread out as they pass through a small opening).

Waves can transport energy and information from one place to another through a medium, but the medium itself is not transported. A disturbance, or change in some physical quantity, is passed along from point to point as the wave propagates. The medium, however, reverts to its undisturbed state after the wave has passed. Waves of different types propagate with widely varying speeds. Light waves and radio waves in space travel 3×10^8 meters (186,000 miles) in one second, for example, whereas sound waves travel only 343 meters (1125 feet) per second in air. Water waves are still slower, traveling only a few feet in a second. Light waves can travel millions of miles through empty space, whereas sound waves require some material medium (gas, liquid, or solid) for propagation.

3.2. Hearing Sound

In a sound wave there are extremely small periodic variations in pressure to which our ears respond in a rather complex manner. The minimum pressure fluctuation to which the ear can respond is less than one billionth (10^{-9}) of atmospheric pressure. This threshold of audibility, which varies from person to person, corresponds to a sound pressure change of about 2×10^{-5} pascals (or newtons/meter^2) at a frequency of 1000 Hz or vibrations per second. The threshold of pain corresponds to a pressure approximately one million (10^6) times greater, but is still less than 1/1000 of atmospheric pressure.

We generally expresses sound pressure level (SPL) on a *decibel* scale. The zero on this scale is at 2×10^{-5} pascals, the nominal threshold of hearing at 1000 Hz (about C_6 or "high C") under ideal quiet conditions. Doubling the sound pressure results in an increase in SPL of 6 dB (decibels). Increasing the sound pressure 10 times results in a 20 dB increase in SPL. Typical sound levels one might encounter are shown in the Table 3.1.

Table 3.1. Typical Sound Levels One Might Encounter

Jet takeoff (60 m)	120 dB	
Construction site	110 dB	*Intolerable*
Shout (1.5 m)	100 dB	
Heavy truck (15 m)	90 dB	*Very noisy*
Urban street	80 dB	
Automobile interior	70 dB	*Noisy*
Normal conversation (1 m)	60 dB	
Office, classroom	50 dB	*Moderate*
Living room	40 dB	
Bedroom at night	30 dB	*Quiet*
Broadcast studio	20 dB	
Rustling leaves	10 dB	*Barely audible*
	0 dB	

3.3. Loudness and Musical Dynamics

Unlike sound pressure level (SPL), which can be measured on a sound level meter, *loudness* is a subjective quality. Two persons hearing the same sound will not always agree on how loud it is. Loudness depends upon the frequency of the sound; sounds in the range of 1000 to 4000 Hz, where our hearing is very sensitive, will nearly always sound louder than sounds of low frequency (or very high frequency) having the same SPL. As we age, our ears become less sensitive to sounds of high frequency, a condition that is known as *presbycusis*.

The approximate range of frequency and sound level in musical performance is compared to the total range of hearing in Fig. 3.1. Note that music rarely approaches the threshold of audibility or the threshold of pain. Nevertheless it makes use of a large dynamic range (approximately 70 dB).

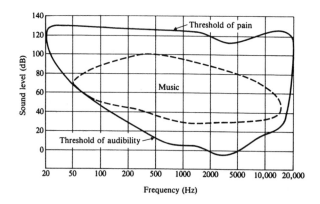

Fig. 3.1. Approximate range of frequency and sound level of music compared to the total range of hearing.

Scientists express subjective loudness in *sones*. Doubling the loudness (in sones) corresponds roughly to a 10 dB increase in SPL. Composers, on the other hand, use dynamic symbols to indicate the appropriate loudness to the performer. The six standard levels are shown in Table 3.2.

Table 3.2. Standard levels of Musical Dynamics

Name	Symbol	Meaning
fortissimo	ff	Very loud
forte	f	Loud
mezzo forte	mf	Moderately loud
mezzo piano	mp	Moderately soft
piano	p	Soft
pianissimo	pp	Very soft

Measurements of SPL of a number of instrumentalists have shown that seldom do musical performers actually play at as many as six distinguishable dynamic levels. Dynamic ranges of wind and string instruments rarely exceed 20 dB, although percussion instruments often play over larger dynamic ranges.

3.4. Sound Power Level

The sound pressure level (SPL) due to a musical instrument obviously depends upon how far the listener is from the instrument. For example, in a free field (outdoors away from reflecting surfaces or in a special anechoic room), the SPL decreases 6 dB for each doubling of the distance. In order to estimate the SPL due to an instrument, it is important to know the source power of the instrument.

Acoustic power, like electrical power, is measured in watts. However, it is customary to express the *power level* (PWL) in decibels by using a suitable reference level, a power of 10^{-12} watts. A bass drum is capable of radiating the most power of all the instruments in an orchestra. The peak acoustical power of a bass drum has been meaured to be about 20 W, which would correspond to a sound power level (PWL) of 133 dB. At a distance of 1 meter in a free field, the sound pressure level (SPL) would be a very loud 122 dB, while at 10 meters, it would still be 102 dB. (The sound pressure level at 1 meter is about 11 dB less than the acoustic power level of the source; see p. 88 in ref. [1]). No wonder a bass drum can be heard regardless of how loud the orchestra is playing!

3.5. Masking Sounds

When the ear is exposed to two or more different tones, it is a common experience that one may mask the others. Masking is probably best explained as an upward shift in the hearing threshold of the weaker tone by the stronger one, and it depends upon the frequencies of the two tones. Masking is generally observed when the tones occur simultaneously, but it can also occur when the tones are not simultaneous. Several demonstrations of masking phenomena are included in the compact disc *Auditory Demonstrations* [2].

From the many masking experiments reported in the literature, we mention some interesting results that apply to hearing the sounds of percussion instruments:
1. Tones close together in frequency mask each other more than tones widely separated in frequency;
2. A pure tone masks tones of higher frequency more effectively than tones of lower frequency.
3. The greater the intensity of the masking tone, the broader the range of frequencies it can mask.
4. A tone can be masked by a sound that ends a short time (up to about 20 or 30 milliseconds) before the tone begins. This is called *forward masking* and suggests that recently stimulated cells are not as sensitive as fully rested cells.
5. A tone can be masked by sound that begins up to ten milliseconds *later*. This surprising result is called *backward masking* and suggests that processing of sound in the brain can be interrupted by another later signal.
6. Under certain conditions a tone in one ear can be masked by a sound in the other ear; this is called *central masking*.

Masking of musical percussion sounds has not been studied very extensively, to my knowledge, but percussive sounds are generally more difficult to mask than steady tones. The loud low-frequency sounds of a bass drum, on the other hand, are able to mask almost any instrument in the orchestra, a fact that overzealous percussionists should keep in mind.

3.6. Loudness and Duration

How does the loudness of an impulsive sound compare to the loudness of a steady sound at the same sound level? Experiments have pretty well established that the ear averages sound energy over about 200 milliseconds, so loudness grows with duration up to this value. Stated another way, loudness level increases by 10 dB when the duration is increased by a factor of 10. The loudness level of broadband noise seems to depend somewhat more strongly on stimulus duration than the loudness level of pure tones, however. Fig. 3.2 shows approximately how loudness level changes with duration.

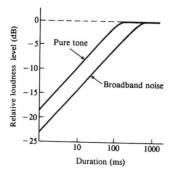

Fig. 3.2. Variation of loudness level with duration [3].

The ear, being a very sensitive receiver of sounds needs some protection to avoid injury by very loud sounds, such as the sound of a bass drum, a tam tam, or a loud cymbal crash. Up to 20 dB of effective protection is provided by muscles attached to the eardrum and the bones in the middle ear (ossicles). When the ear is exposed to sounds in excess of 85 dB or so, these muscles tighten the ossicular chain and pull the stapes (stirrup-shaped bone) away from the oval window of the cochlea. This action is termed the *acoustic reflex*.

Unfortunately the reflex does not begin until 30 or 40 ms after the sound overload occurs, and full protection does not occur for another 150 ms or so. In the case of a loud impulsive sound, this is too late to prevent injury to the ear. It is interesting to speculate what type of protective mechanism, analogous to eyelids, might have developed in the auditory system had the loud sounds of the modern world existed for millions of years (earlids, perhaps?). It is well known that many musicians (orchestral as well as jazz and rock) suffer from hearing loss due to prolonged exposure to high sound levels.

References

1. T. D. Rossing, *The Science of Sound* (Addison-Wesley, Reading, MA, 1990).
2. A. J. M. Houtsma, T. D. Rossing, and W. M. Wagenaars, *Auditory Demonstrations* (Philips Compact Disc #1126-061 and text, 1987).
3. J. J. Zwislocki, *J. Acoust. Soc. Am.* **46** (1969) 431.

Chapter 4
Drums with Indefinite Pitch

In this chapter we will consider drums that do not convey a definite pitch, or at least convey a much weaker sense of pitch than the drums considered in Chapter 2. Most drums in common use fall into this category.

4.1. Tom Toms

Originally describing an African drum, the term *tom tom* has now come to denote side drums of different sizes and shapes having one or two heads, the more indefinite pitch of the two-headed type usually being preferred for orchestral use. Head diameters vary from 20 to 45 cm (8 to 18 in) in diameter.

Drummers often add a circular patch or dot of material to increase the thickness of the head at the center. These are said to give a "centered," slightly "tubby" sound with some semblance of pitch. The probable reason for this observation can be seen from Table 4.1 which gives modal frequencies of a 30-cm diameter single-head tom tom without a dot and with various size dots added to the membrane. Without the dot, the modes are quite inharmonic, but adding the dot brings them somewhat closer to a harmonic relationship.

Table 4.1. Modal frequencies of a 30-cm diameter tom tom without and with center dots [1]

Mode	No dot f	f/f_{01}	8.9-cm dot f	f/f_{01}	11.4-cm dot f	f/f_{01}	14-cm dot f	f/f_{01}
0,1	147	1	144	1	142	1	140	1
1,1	318	2.16	307	2.13	305	2.15	289	2.06
2,1	461	3.14	455	3.16	450	3.17	434	3.10
0,2	526	3.58	522	3.63	485	3.42	474	3.39
3,1	591	4.02	583	4.05	581	4.09	566	4.04
1,2	684	4.65			681	4.80		
4,1	703	4.78	693	4.81	701	4.94	693	4.95

4.2. Interlude: Pitch Glide in Membranes

The vibration frequency of a membrane increases when the tension increases (like a vibrating string, the frequency is proportional to the square root of tension: see Section 2.2). When a membrane vibrates with very large amplitude, the average tension increases and hence the frequency increases. Thus the pitch of a drum drops slightly as the vibration amplitude decreases during the sound decay. This is an example of a nonlinear effect (so-called because a graph of deflection *vs* applied force is a curved line rather than a straight line). This pitch glide is familiar in the sound of most drums, and jazz drummers make use of it for special effects.

While doing research for his master's thesis at Northern Illinois University, Cal Rose

Drums with Indefinite Pitch

discovered that pitch glide in a tom tom can be greatly enhanced by adding a Mylar ring to load the outer portion of the drumhead. He traced this enhanced effect to the added stiffness to shear in the thicker membrane. Cal patented the idea and has since supplied "Tonga rings" to a number of drummers. The dependence of frequency on amplitude in a 33-cm tom tom is shown in Fig. 4.2. Note the slight curvature (nonlinearity) in the uniform membrane and the much greater curvature in the head with an annular ring.

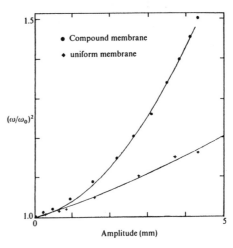

Fig. 4.2. Dependence of frequency on amplitude in a 33-cm tom tom head vibrating in its fundamental (01) mode. The vertical axis is the square of the ratio of frequency to the low-amplitude frequency. The tension was set near the low end of the playing range [2].

4.3. Interlude: Modes of a Two-Mass Vibrator

The vibrations of a two-headed tom tom and other two-headed drums, such as snare drums and bass drums, are somewhat more complex than those of a drum with one head, because there is both acoustical and mechanical interaction between the two heads. In order to understand this interaction, it is worthwhile taking time to observe what happens in a simple vibrating system consisting of two masses attached to three springs, as shown in Fig. 4.3.
The system will have two normal modes of vibration, one in which the masses move in the same direction and another (at a higher frequency) in which they move in opposite directions.

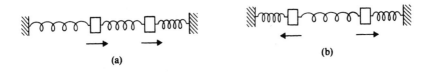

Fig. 4.3. Normal modes of a two-mass vibrator. In the mode of lower frequency, the masses move in the same direction, but in the mode of higher frequency, they move in opposite directions.

We will apply this principle to two-headed drums. The two outer springs will correspond to the tensions in the two membranes, while the center spring corresponds to the elasticity of the enclosed air and also the drum shell which couples the two heads.

4.4. Snare Drum

Snare drums or side drums have evolved over several centuries. The modern orchestral snare drum is a two-headed instrument about 35 cm (14 in) in diameter and 13-20 cm (5-8 in) deep. When the upper or *batter* head is struck, the lower or *snare* head vibrates against strands or cables of wire or gut (the snares). Alternately, the snares can be moved away from the lower head to give a totally different sound, more like that of a tom tom.

There is appreciable coupling between the two heads of a snare drum, especially at the lower frequencies. This coupling can take place acoustically, through the enclosed air, or mechanically, by way of the drum shell, and it leads to pairs of modes (see section 4.3). A pair of modes based on the (01) membrane modes, for example, are shown in Fig. 4.4a. In the pair member with the lower frequency, the two heads move in the same direction, while in the member of higher frequency, the two heads move in opposite directions (compare Fig. 4.3).

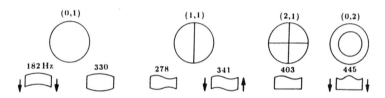

Fig. 4.4. The six lowest resonances observed in a snare drum include two mode pairs based on the (01) and (11) membrane modes.

In the (11) mode pair, the member in which the drum heads move in opposite directions has the lower frequency. This is due to the added mass of the enclosed air that "sloshes" from side to side. In the higher (11) mode, the air moves a smaller distance, essentially normal to the plane of the heads, and the mass loading is diminished. Arrows in Fig. 4.4 indicate the modes in which appreciable motion of the drum shell takes place.

A simple two-mass model describes the first mode pair reasonably well [3]. Each head is represented by a mass and a spring, and the enclosed air constitutes a third spring connecting the masses. Generally the batter head and snare heads have different thickness (mass), and they are tuned to different frequencies (tensions). In the lower member of the pair, the shell moves opposite to the two heads, and so the mode frequency will depend upon the mass of the shell and the way in which the drum is supported. A drum placed on a drum stand, for example, sounds quite different from the same drum on a rigid support or when freely supported by springs or elastic rope. Drums with heavy metal shells generally sound different from drums with less massive shells.

Frequencies of some of the vibrational modes observed in a 36-cm diameter snare drum are given in Table 4.2. In order to measure mode frequencies in each head alone, the opposite head was damped with sandbags; thus, the frequencies are for each drumhead backed by the enclosed air.

Table 4.2. Mode frequencies in a 36-cm snare drum [4]

Mode	Batter head (Hz)	Snare head (Hz)	Drum (Hz)
0,1	227	299	182, 330
1,1	284	331	278, 341
2,1	403	507	403
0,2	445	616	442
3,1	513	674	512
1,2	555	582	556
4,1	619	859	619

Shell Vibrations

The lowest modes of the free drum shell are the cylindrical shell modes having m nodal lines parallel to the axis and n circular nodes, as shown in Fig. 4.5a. Holographic interferograms of six of these modes are shown in Fig. 4.5b, in which the nodal lines appear as white lines. The outside and inside surfaces were recorded simultaneously by using two object beams.

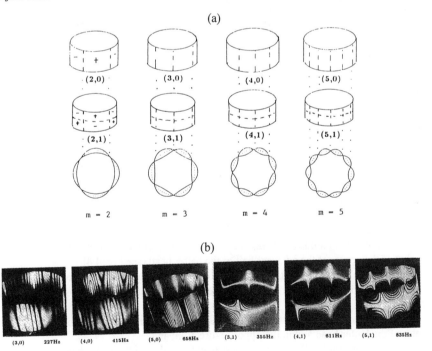

Fig. 4.5. (a) Vibrational modes of a free drum shell; (b) Holographic interferograms of six modes [3].

Shell motion in the complete drum is considerably more complicated than in the free drum shell. In addition to the up and down motion of the shell, shown in Fig. 4.4, there is radial motion driven by the vibrations of the drum heads and by the motion of the enclosed air. In most vibrational modes of the drum, shell motion is much less than the motion of the drum heads. In a few modes, however, shell motion is appreciable. Some of these modes are shown holographically in Fig. 4.6.

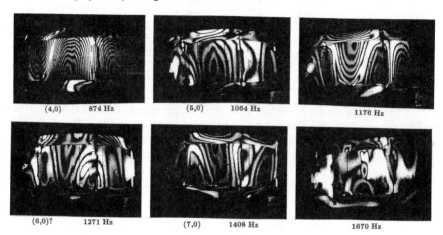

Fig. 4.6. Holographic interferograms of the complete drum showing modes of vibration in which shell motion is appreciable. Shell motion resembles the modes of the free drum shell in Fig. 4.5 [3].

In one experiment on shell motion, the vertical motion of the drum, when freely suspended on rubber bands, was determined by attaching an accelerometer at eight different locations around the circumference of the ring while the drum was driven at the resonance frequencies of several modes. The acceleration amplitudes were compared to those of the batter head. The following results were obtained [3]:

1. In the parallel (01) mode (181 Hz), the shell moved oppositely to the heads with an amplitude approximately 1% of that of the batter head;
2. In the antiparallel (01) mode (322 Hz), the shell amplitude was less than 0.2% of that of the batter head;
3. In the parallel (11) mode (340 Hz), the shell "rocked" with an amplitude approximately 1% that of the center of the adjacent half of the batter head (see Fig. 4.4b);
4. In the antiparallel (11) mode (272 Hz), the shell rocking amplitude was about 0.3% of that of the batter head, and it moved in the same direction as the batter head (opposite to the snare head);
5. In the (02) mode of the batter head (445 Hz), the shell amplitude is about 0.5% of the center of the batter head and in the opposite direction; and
6. In other modes, the shell motion was less than 0.2% that of the batter head.

Drums with Indefinite Pitch

Sound radiation from the drum

Sound radiation patterns in the vertical plane for the four lowest modes of a snare drum are shown in Fig. 4.7. The pattern for the lowest mode [the parallel (01) mode in Fig. 4.7a] has a strong dipole component (see Section 2.6), as expected (since one head is moving in when the other head is moving out), with maximum radiation along the center axis of the drum. The sound pressure level in front of the batter head is about 2 dB greater than in front of the snare head but 13 dB greater than at the mid-plane of the drum. The radiation from the other (01) mode [the antiparallel (01) or "breathing" mode in Fig. 4.7b] is considerably weaker (for the same driving force) and is nearly the same in all directions.

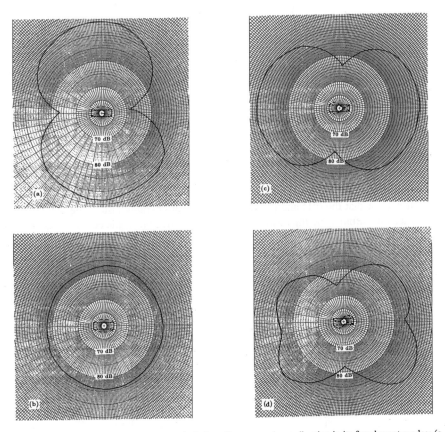

Fig. 4.7. Sound radiation patterns in the vertical plane for a snare drum vibrating in its four lowest modes: (a, b) (01) modes at 182 Hz and 330 Hz; (c,d) (11) modes at 278 Hz and 341 Hz [4].

Note that the patterns in Fig. 4.7 (a) and (c) show a dipole character, while the pattern in Fig. 4.7 (d) has a marked quadrupole pattern, as expected, since each drum head acts as a dipole source. Sound radiation patterns in the horizontal plane from the (11) mode pair are shown in Fig. 4.8. It is interesting to compare these to those of the single-headed timpani in Fig. 2.8.

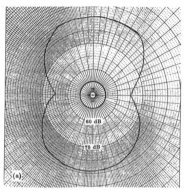 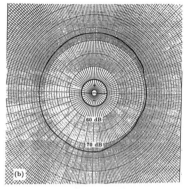

Fig. 4.8. Sound radiation patterns in the horizontal plane for a snare drum vibrating in its two (11) modes: (a) 278 Hz; (b) 341 Hz [4].

Modal Decay Rates

Although energy loss in a snare drum is mainly due to radiated sound, mechanical loss of vibrational energy through the drum shell to its supports is not negligible. Modal decay rates for several modes in a drum tuned to a relatively low tension are given in Table 4.3. It is clear that the (01) and (02) modes decay considerably faster when the drum is supported on a drum stand than when it is freely supported on elastic cords. These modes incorporate an appreciable amount of vertical shell motion, and so vibrations are transmitted to the supports. Decay rates of other modes depend much less on the type of support.

Table 4.3. Modal decay rates (in dB/s) of a snare drum freely supported by elastic cords and supported on a drum stand. The drum was struck at the center of the batter head (columns 1 and 3) or half way between the center and the rim (columns 2 and 4) [3].

Mode	Elastic cords		Drum stand	
	Center	Off-center	Center	Off-center
01	35	38	70	60
11	40	30	40	30
11	38	49	47	...
01	45	25	47	30
21	...	...	24	30
02	40	30	65	50
31	22	32	...	35
41	47	45	58	65

Drums with Indefinite Pitch

Snare action

The coupling between the snares and the snare head depends upon the mass and the tension of the snares as well as the head. At a sufficiently large amplitude of the snare head, properly adjusted snares will leave the head at some point during the vibration cycle and then return to strike it, thus giving the snare drum its characteristic sound. The larger the tension on the snares, the larger the amplitude needed for this action to take place.

Velocities of the snare head and snares in one snare are shown in Fig. 4.9. The snare velocity initially follows a simple sine curve whose period is greater than that of the snare head. Therefore, the snare head reverses its direction first, and the snares lose contact with the head. The smooth snare curve is disturbed when the snares, vibrating back ($v_s<0$), meet the head which is already moving in the opposite direction ($v_h>0$). Through the impact, higher modes of vibration are excited in the snares as well as the snare head.

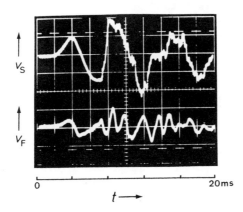

Fig. 4.9. Velocity waveforms for snares (v_s) and snare head (v_h) in a snare drum. Arrow marks the time at which snares first strike the head [5].

For the snares to sound at all requires a certain amplitude of the snare head. This critical amplitude increases with the snare tension. The snare tension is optimum when both the head and the snares are moving at maximum speed in opposite directions at the moment of contact. In this case, the impact (and the radiated sound) is the greatest.

Figure 4.10 shows sound spectra for a drum with three blow strengths and two snare tensions. At low snare tension, the medium blow causes the head to exceed its critical amplitude (Fig. 4.10b), whereas at the higher tension the strongest blow is required (Fig. 4.10f). Note that the damping effect of the snares, as shown by a broadening of the fundamental head resonance peak, is greater at low snare tension (Fig. 4.10a).

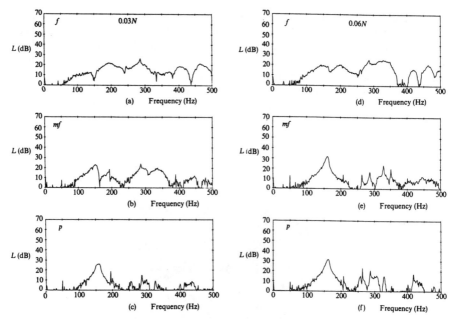

Fig. 4.10. Sound spectrum of a snare drum for three blow strengths and two values of snare tension (0.03 Newtons and 0.06 Newtons) [5].

4.5. Bass Drum

The bass drum is capable of radiating the most power of all the instruments in the orchestra (A peak acoustical power of 20 W was observed by Sivian et al [6]). A concert bass drum usually has a diameter of 80-100 cm (32-40 in), although smaller drums (50-75 cm or 20-30 in) are popular in marching bands. Most bass drums have two heads, set at different tensions, but single-headed "gong" drums are widely used when a more defined pitch is appropriate. Mylar heads with a thickness of 0.25 mm (0.010 in) are widely used, although calfskin heads are preferred by some percussionists for large concert bass drums.

Most drummers tune the *batter* or beating head to a greater tension than the *carry* or resonating head; some percussionists suggest that the difference be as much as 75%, equivalent to an interval of about a musical fourth [7]. Levine [8] recommends tuning the carry head higher than the batter head for orchestral music but lower than the batter head in solo or small ensemble playing. A distinctive timbre results from setting both heads at the same tension, but the prominent partials in the 70-300 Hz range appear to be stronger (initially) and to decay faster when the carry head is tuned below the batter head [9].

Modal frequencies of an 82-cm (32-in) diameter bass drum are given in Table 4.4.

Frequencies of the (01), (11), (21), (31), and (41) modes fall surprisingly near a harmonic series, and if their partials were the only ones heard, the bass drum would be expected to have a rather definite pitch. In the frequency range above 200 Hz, however, there are many inharmonic partials that probably sound louder since the ear is less sensitive at low frequency (see Fig. 3.1). Fletcher and Bassett [10] found 160 partials in the frequency range 200-1100 Hz, although their scheme for associating these partials with vibrational modes of the membrane does not appear to be correct [9]).

Table 4.4. Modal frequencies observed in the batter head of an 82-cm (32-in) diameter bass drum [9].

Mode	Carry head at lower tension	Heads at same tension
0,1	39	44, 104
1,1	80	76, 82
2,1	121	120
3,1	162	160
4,1	204	198
5,1	248	240

Coupling between the two heads gives rise to mode pairs, just as in a snare drum. When the two heads in one drum were at the same tension, for example, a pair of (01) modes was observed at 44 Hz and 104 Hz and a pair of (11) modes at 76 Hz and 82 Hz. Setting the two heads to the same tension tends to maximize this coupling.

Removing the carry head changes the modal frequencies but little from the values in the first column in Table 4.4 (for the carry head tuned below the batter head). Decay rates vary from 3 to 9 dB/s when the carry head is at a lower tension than the batter head; this increases to 6-11 dB/s when the heads are at the same tension. Removing the carry head gives decay rates from 3 to 8 dB/s, about the same as in the preferred playing arrangement of unequal tensions in the two heads.

A rather substantial frequency shift due to an increase in average tension is observed when the bass drum is struck a full blow. Cahoon [11] has observed an upward frequency shift of 10%, nearly a whole tone on the musical scale, when the initial amplitude is 6 mm (¼ in) However, the corresponding pitch shift is less, since most listeners experience a downward pitch shift with intensity at low frequency (see, for example Chap. 7 in ref. 12).

4.6. Conga Drums

Conga drums or tumbas are large Latin American hand drums that probably evolved from African drums made from hollowed-out trees. Modern conga drums are generally constructed of long strips of wood, grooved, and glued together or of synthetic material, and a tensioning mechanism is added to tune the heads. Head diameters range from 23 to 30 cm (9-12 in), and their fundamental frequencies are in the range of 131 to 262 Hz (C_3 to C_4). The individual drums are sometimes designated "quinto," "conga," and "tumba" (from small to large, respectively).

There are three basic sounds on the conga:
1. An open tone produced by striking the rim with the palm of the hand allowing the fingers to slap on the head;
2. A slap produced by striking the rim with the palm of the hand near the rim, allowing the fingers to be slapped onto the head;
3. A bass tone, in which the fleshy part of the hand strikes the center of the head and rebounds, allowing the head to vibrate freely in its fundamental mode;
4. A glissando created by the pressure of the fingertips sliding across the drum head.
Other important sounds are called the "wave" (or heel-and-toe) in which the striking hand makes a rocking motion, and the closed tone, produced by striking the drum at or near the center with the palm of the hand, which remains in contact long enough to damp the head.

Sound spectra of the open tone, slap, and bass tone are shown in Fig. 4.11. The first two peaks around 80 Hz and 240 Hz appear to be caused by (01)-type motion of the head, coupled to the first two pipe-like modes of the tubular body. The next peak, around 420 Hz (not present in the bass tone since the drum is struck at the center), corresponds to the (11) mode in the head. The peak at 620 Hz probably results from the (21) mode, but we have not verified this in the laboratory [13].

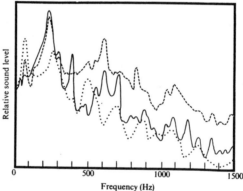

Fig. 4.11. Sound spectra for three different strokes on a conga drum: open tone —, slap - - -, and bass tone ••

4.7. Bongos and Timbales

Bongo drums are the smallest and highest pitched of the Latin American drum family. They are played in pairs, and range from about 15 cm to 25 cm (6 to 10 in) in diameter. Their wooden shells are slightly conical and open at the bottom. In the original version, the skins (usually goat skins) were nailed to the shell, but most modern bongos have tensioning screws. There is usually an interval of about a musical fourth between the pitch of the two drums.

The player generally holds the drums between the knees, with the larger drum on the right. The fingers and hands are used to produce a variety of both open, ringing sounds and muffled sounds. The sound produced is high and penetrating.

Timbales are the middle range of the Latin American drum family. They were originally small kettle drums, consisting of animal hide stretched over wooden bowls, but modern timbales generally have metal shells 30 to 35 cm (12 to 14 in) in diameter. The shells are either totally open at the bottom or they have a sound hole about the size of a hand. For Latin American music the heads are tightly tensioned and the timbales are played with light wooden sticks. A variety of sounds are obtained by hitting the head, the rim, or the outside shell of the drum.

Conga drums, bongos, and timbales are shown in Fig. 4.12.

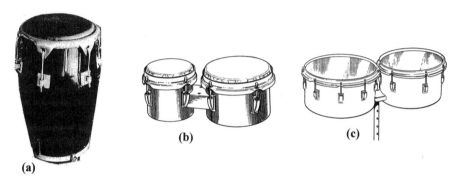

Fig. 4.12. Latin American drums: (a) conga drums; (b) bongos; (c) timbales.

4.8. Rototoms

The rototom is a single-membrane tunable drum that has become very popular in schools, marching bands, and rock bands. They were apparently invented by the American percussionist-composer Michael Colgrass, and they are now manufactured by the Remo company. They are essentially a tunable tom-tom without a shell. The counter hoop is connected by a light alloy frame to a center spindle, so that by turning the drum clockwise or counter-clockwise the tension on the drumhead is increased or decreased.

Their portability, size, and relatively low cost have made them very popular instruments. Yet, they have a very clear resonant sound, and are useful for high timpani notes in an orchestra. A glissando is obtained by turning the drum with one hand while playing it with the other. Seven sizes, from 6 to 18 inches in diameter, cover a 3-octave range from E_2 to E_5. In spite of their popularity, no discussion of their acoustical properties appear to have been published in either the acoustical or musical literature.

The sound spectrum of a 10-inch rototom tuned to about the middle of its range is shown in Fig. 4.13. The spectrum is dominated by radiation from the fundamental (0,1) membrane mode having no nodal diameters and one nodal circle at the edge. Other prominent modes are the (3,1) mode (having 3 nodal diameters), at 561 Hz, the (0,2) mode (with two nodal circles) at 370 Hz, and the (1,1) mode at 310 Hz.[14]

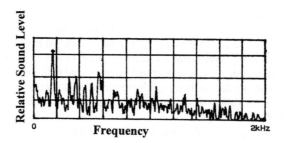

Fig. 4.13. Sound spectrum of a 10-inch diameter rototom. Vertical scale is 20 dB/division [14].

The frequency ratios of some of the principal modes to the fundamental (0,1) mode frequency are shown in Fig. 4.14 for a wide range of tension (approximately 6.5 times). The playing range would generally be considered to be about the center half of this total tension range. Note that the ratios vary somewhat over the playing range, but not enough to cause a large change in timbre.

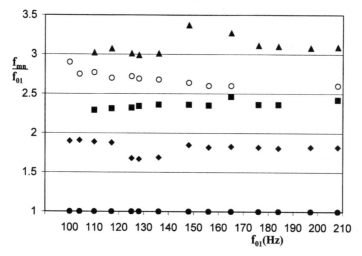

Fig. 4.14. Mode frequencies of a 10-inch rototom at different membrane tensions compared to the fundamental (0,1) mode: f_{01} ●; f_{11} ◆; f_{02} ■; f_{21} ○; f_{31} ▲   [14].

By means of electronic TV holography, we have recorded mode shapes for the main resonances of the drum. The holographic system used for these measurements has been described previously [1,2]. Mode shapes for eight modes are shown in Fig. 4.15.

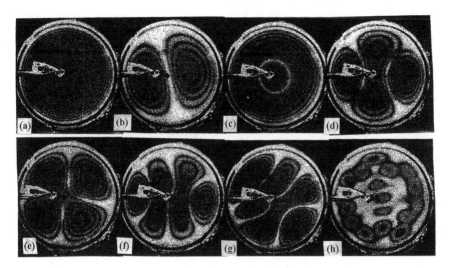

Fig. 4.15. Membrane mode shapes in a 10-inch rototom recorded using electronic TV holography. (a) (0,1) mode at 168 Hz; (b) (1,1) mode at 310 Hz; (c) (0,2) mode at 370 Hz; (d) (2,1) mode at 440 Hz; (e) (2,1) mode at 448 Hz; (f, g) (3,1) mode with some of the (1,2) mode mixed in at 561 Hz and 589 Hz; (h) configuration at 1055 Hz including the (7,1) mode [14].

The modal shapes observed in rototoms may be compared to those observed in tom toms (section 4.1) and snare drums (section 4.4), drums of comparable construction. In all four drums, (0,1), (1,1), (2,1), (3,1) and (0,2) modes are observed, although in slightly different orders of frequency. In the two-headed snare drum, coupling between the two heads and the enclosed air leads to (0,1) and (1,1) mode pairs. The (1,2) mode was not observed as a separate mode in the rototom, but its effect is seen in the resonances observed at 561 and 589 Hz, which appear to be due to the (3,1) mode with a little of the (1,2) mode mixed in.

The 10-inch rototom head has a center patch of Mylar 3½ inch in diameter, but this doesn't seem to affect the mode shapes, which are essentially those of a uniform membrane. Mode frequencies and the coupling between modes may be affected by the patch.

The rototom sounds a strong clear tone with an easily identifiable pitch corresponding to the fundamental. Ordinarily a clear sense of pitch requires one or more harmonic partials in the sound, but this does not appear to be the case in the rototom.

In the large membrane of a timpani, the fundamental (0,1) mode decays very rapidly due to radiation damping, and striking the membrane near the center produces a rather dull "thud." When played, a timpani is typically struck about half way to the rim, which strongly excites the harmonically tuned (1,1), (2,1), and (3,1) modes [5].

4.9. Irish Bodhrán

The traditional Irish *bodhrán* (pronounced "bow-rawn") is a single-headed frame drum about 18 inches in diameter. The rim, which is generally constructed of ash, beech, or birch, is 2 to 6 inches deep, and the drum is reinforced by one or two crossbraces. It can be played with either the hands or a small stick, as shown in Fig. 4.16.

The bodhrán was traditionally played while standing, dancing, or marching but players today are usually seated. The drum is rested on the left knee and struck with the right hand using a loose wrist movement that has been compared to shaking water off one's hands. A glissando effect may be achieved by applying pressure with the left hand on the inside of the head. [15]

Fig. 4.16. Irish *bodhrán* [15].

4.10. African Drums

Drumming in Africa is an integral part of most religious and secular ceremonies. Drums are played in traditional rituals, in worship ceremonies held in churches and mosques, in village squares, town clubs and wherever people mingle. Drumming has long been such an important part of music and culture in Africa, it is not surprising that hundreds of different drums are played in various countries. It would be impractical to attempt to list them all, much less to describe their acoustical properties. We will cite only a few examples.

Paging through a book on African drums, such as *Drums, the Heartbeat of Africa* [18], one notes that many old and new drums can be described as kettledrums, goblet-shaped drums, or hour-glass drums (whose tension can be changed while playing). Examples of each type will be described briefly.

Goblet-shaped Drums

Goblet drums consist of a single membrane stretched over a more-or-less cylindrical upper chamber with an open conical stem below. They are so common throughout Africa that the term "African drum" is often used to describe goblet-shaped drums.

One of the most popular African goblet-shaped drums in Western countries is the *djembé*, most often associated with the Malinke people of Mali and Guineé but played in other African countries as well. The membrane is most often goatskin, stretched and fastened by complex lacework to a wooden body. Rattles, tin sheets (sèssè), and wire rings may be attached to the edge of the skin. The djembé, played only with the hands, is capable of producing a wide variety of sounds. The sound includes a strong "bass" note around 70 to 80 Hz, which appears to be due to the Helmholtz resonance of the shell. In one djembe, we observed the (0,1) mode of the head at 351 Hz, and a pair of (1,1) modes (with the nodal line in two different directions) at 517 and 549 Hz. The (2,1) mode was at 741 Hz and the (3,1) mode at 970 Hz. Partials in the djembe sound have been observed out to 3000 Hz or more [22].

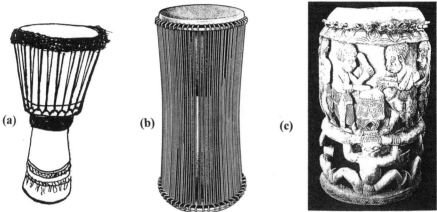

Fig. 4.17. (a) Djembé with goblet shape; (b) Talking drum with hour-glass shape; (c) Carved kettledrum from Cameroon.

Talking Drums

One of the most important aspects of instrumental music in Nigeria and other parts of Africa is the use of musical instruments to speak or talk. Talking instruments are particularly common among Nigerian ethnic groups whose languages are tone languages. In a tone language, a word can have several meanings, depending the tone level or combination of tone levels with which the word is pronounced, which enables musical instruments to act as speech surrogates.

Drums with an hour-glass shape, having tension straps tied around the drum shell, are especially useful as talking drums. One way to play a talking drum is to hold it under the armpit so that the underarm can press or release the tensioning straps to change the pitch when it is struck by a stick held in the opposite hand. Another way is to hang the drum over a shoulder so that it can be pressed to the thigh or the hip to change the tension.

Kettledrums

African kettledrums consist of one or two membranes stretched over kettles of nearly every size and shape imaginable. Often they are intricately carved with figures that represent the primary use of the drum, such as the drum in Fig. 4.17c from Cameroon. Male and female figures are favorite subjects of carvings on drums.

4.11. Japanese Drums

Drums have been used for centuries in Japanese temples. In Buddhist temples, it has been said that the sound of the drum is the voice of Buddha. In Shinto temples it is said that drums have a spirit (*kami*) and that with a drum one can talk to the spirits of animals, water and fire. Drums were also used to motivate warriors into battle and to entertain at town festivals and weddings [17].

More recently, the Japanese *taiko* (drum) has come out of its traditional setting, and today's taiko bands have given new life to this old tradition. Japan's most famous taiko band, the Kodo Drummers, who represent the pinnacle of taiko drumming, have performed in many countries of the world.

Japanese drums fall into two main classes: braced heads and nailed heads. In the nailed drums, the drumheads are fastened to convex barrel-shaped cylinders of wood. In braced drums, the drum heads are secured to iron rings, larger in diameter than the supporting cylinders. The two heads are connected by a cord that passes in "W" form through holes along the edges of the two heads.

Nailed drums include the large o-daiko, the turi-daiko ("hanging drum"), and the ko-daiko. [*Taiko* is the generic name for drum, any prefix turning the initial *t* to *d*, as in *o-daiko* (large drum)]. Braced drums include the da-daiko, the ni-daiko, the kakko, the uta-daiko and the tsuzumi. [The term *tsuzumi* also means drum, but it is generally applied only to braced drums].

O-Daiko

The *o-daiko* is a large nailed drum consisting of two cowhide membranes stretched tightly across the ends of a wooden cylinder 50-100 cm (20-40 in) in diameter and about 1 m (3 ft) in length. The drum, hanging freely in a wooden frame, is struck with large felt-padded beaters. It is often used in religious functions at shrines, where its deep rumbling sound adds solemnity to the occasion. It occasionally appears in orchestral pieces, such as Orff's "Prometheus" (1968). An o-daiko is shown in Fig. 4.28a.

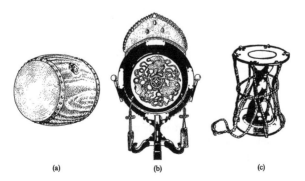

Fig. 4.18. Japanese drums: (a) o-daiko; (b) turi-daiko; and (c) kotodumi.

Obata and Tesima [18] found modes of vibration in the o-daiko to be somewhat similar to those in the bass drum (Table 4.4). Pairs of (0,1) modes were found to have frequencies that ranged from about 168 Hz and 193 Hz, under conditions of high humidity, to 202 Hz and 227 Hz at low humidity. In the higher mode of each pair, the two membranes move in opposite directions. The next mode at 333 Hz was identified as the (1,1) mode, and the (2,1) mode was found at 468 Hz.

Mode frequencies are given in Table 4.5. Since air loading is substantial for a large membrane, it is quite surprising that the observed frequencies are in such good agreement with the theoretical values for an unloaded membrane. This would suggest that the membrane itself has a large mass and a high tension compared with Western drums. The absence of a harmonic relationship between modal frquencies indicates that the drum has an indefinite pitch.

Table 4.5. Modal frequencies of the o-daiko [18]

Mode	Frequency	f/f_{01}	f/f_{01} (ideal)	f/f_{11}	f/f_{11} (ideal)
0,1	202	1.00	1.00	0.61	0.63
0,1	227	1.12		0.68	
1,1	333	1.65	1.59	1.00	1.00
	383	1.90		1.15	
2,1	468	2.32	2.13	1.41	1.34
0,2	492	2.44	2.29	1.48	1.44
3,1	544	2.69	2.65	1.63	1.66
1,2	621	3.07	2.92	1.86	1.83
4,1	695	3.44	3.15	2.09	1.98
	739	3.66		2.22	
	865	4.28		2.60	
	905	4.48		2.72	
	1023	5.06		2.07	

The *turi-daiko*, shown in Fig. 4.18b, is a small hanging drum used in Japanese drama and in the classical orchestra. The cylindrical wooden bodies of these drums are usually hollowed out of a single log. Obata and Tesima [18] measured the modal frequencies of a shallow turi-daiko 29 cm in diameter with a length of 7 cm. A single (0,1) mode was observed to have a frquency of 195 Hz. Their experiments showed that the fundamental frquency is lowered by increasing the length (volume) of the drum.

Tsuzumi or Tudumi

The *tsuzumi* or *tudumi* is a braced drum consisting of a body with cup-shaped ends and leather heads on both ends. There are two types of tsuzumis: o-tsuzumi and ko-tsuzumi. The length of the o-tsuzumi (about 29 cm or 11 in) is typically a little larger than the ko-tsuzumi (25 cm or 10 in), but the main difference is the drumhead. Whereas the o-tsuzumi has heads of uniform density, the top head of the ko-tsuzumi has a circular groove called the *kannyu* which forms a sort of annular membrane having vibrational mode frequencies that are nearly harmonic [19]. A few sheets of Japanese paper wet with saliva (called *choshigami*) cover an area of about 1 cm² at the center of the bottom head, which tunes the modes of this head in a nearly harmonic relationship as well [19].

Professional players use a number of different playing styles in performing on the o-tsuzumi and ko-tsuzumi. *Kashira* is a style of forcefully striking the membrane, while *kan* is a style where the membrane is struck normally. In the *otsu* style of playing the ko-tsuzumi, the player loosens the *shirabeo* just after striking, causing a downward pitch glide; pitch changes of 16% have been noted [19].

Figure 4.19 shows spectra of an o-tsuzumi tone played inthe kashira style and a ko-tsuzumi tone played in the otsu style. The o-tsuzumi tone shows a strong doublet at frequencies of 1717 Hz and 1834 Hz which orignates from the (1,1) mode. The ko-tsuzumi tone, has a strong fundamental of 270 Hz and several other partials closely harmonic in frequency to the fundamental.

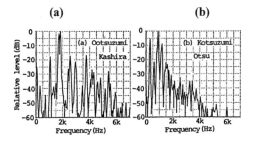

Fig. 4.19. Spectrum of o-tsuzumi played in the kashira style; (b) spectrum of ko-tsuzumi played in the otsu style [19].

Spectra of the o-tsuzumi tones and ko-tsuzumi tones played in several styles are shown in Fig. 4.20. Mode designations and frequency ratios to the fundamental appear above each spectrum. Note the nearly harmonic relationship of partials in the ko-tsuzumi spectrum [19].

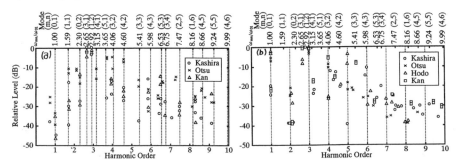

Fig. 4.20. (a) Spectrum of o-tsuzumi played in three styles; (b) spectrum of ko-tsuzumi played in four styles[17].

4.12. Indonesian Drums

Drums are very important instruments in both Balinese and Javanese gamelans. The drummer in a gamelan sets the tempo and thus, in a sense, conducts the gamelan. Nearly all *kendang* or drums in the gamelan are two headed, and they come in various sizes. A large kendang gending and a small kendang ciblon are shown in Fig. 4.21. Note the difference in the cross sections of the carved shells. The sound spectrum of a Balinese kendang is shown in Fig. 4.22. The nearly harmonic overtones give the drum a sense of pitch.

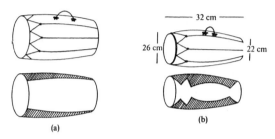

Fig. 4.21. Javanese drums: (a) kendang gending and (b) kendang ciblon [20].

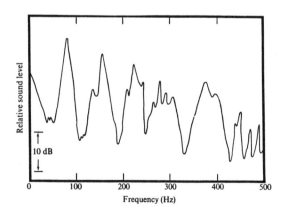

Fig. 4.22. Sound spectrum of a Balinese drum [19].

References

1. T. D. Rossing, *The Physics Teacher* **15** (1977) 278.
2. C. D. Rose, *A New Drumhead Design: An Analysis of the Nonlinear Behavior of a Compound Membrane,* M.S. thesis, Northern Illinois University, 1978.
3. T. D. Rossing, I. Bork, H. Zhao, and D. Fystrom, *J. Acoust. Soc. Am.* **92** (1992) 84.
4. H. Zhao, *Acoustics of Snare Drums: An Experimental Study of the Modes of Vibration, Mode Coupling and Sound Radiation Pattern,* M.S. thesis, Northern Illinois University, 1990.
5. I. Bork, *Entwicklung von akustischen Optimierungsverfahren für Stabspiele und Membraninstrumente,* PTB report, Project 5267, Braunschweig, Germany.
6. L. J. Sivian, H. K. Dunn, and S. D. White, *J. Acoust. Soc. Am.* **2**, 330.
7. J. P. Noonan, *The Instrumentalist* (October 1951) 53.
8. D. Levine, *The Instrumentalist* (June 1978) 545.
9. T. D. Rossing, *J. Acoust. Soc. Am.* **82** (1987) S69.
10. H. Fletcher and I.G. Bassett, *J. Acoust. Soc. Am.* **64** (1978) 1570.
11. D. E. Cahoon, *Frequency-oTime Analysis of the Bass Drum Sound,* M.S. thesis, Brigham Young University.
12. T. D. Rossing, *The Science of Sound,* 2nd ed. (Addison-Wesley, Reading, MA, 1990).
13. N. H. Fletcher and T. D. Rossing, *The Physics of Musical Instruments,* 2nd ed. (Springer-Verlag, New York, 1998).
14. T. D. Rossing and J. Kwon, *J. Acoust. Soc. Am.* **108**, (2000).
15. T. O'Mahoney, *Percussive Notes* **37**(2), 34-40 (1999).
16. E. A. Dagan, ed., *Drums, the Heartbeat of Africa* (Galerie Amrad African Art publications, Montréal, 1993).
17. M. Gould, *Percussive Notes* (April 1996) 41.
18. J. Obata and T. Tesima, *J. Acoust. Soc. Am.* **6** (1935) 267.
19. S. Ando, Paper 4aMUb3, Acoust. Soc. Am./Acoust. Soc. Japan, Honolulu, 1996.
20. J. Lindsay, *Javanese Gamelan* (Oxford, Singapore, 1979).
21. T. D. Rossing and R. B. Shepherd, *Percussive Notes* **19**(3) (1982) 73.
22. Albert Prak <www.drums.org/djembefaq/v20a.htm>

Chapter 5
Interlude: Vibrations of Bars and Air Columns

Vibrations of strings and membranes were discussed in Chapter 2. We learned that an ideally flexible string fixed at its ends has modes of vibration in which the string is divided into any number segments that move in opposite directions. The oppositely moving segments are separated by *nodes* or pivot points and the mode frequencies are harmonics or multiples of a fundamental mode. The wave speed on such a string is independent of frequency; that is, for a given tension, all transverse waves travel at the same speed.

In a two-dimensional membrane, on the other hand, even an ideally flexible one, the vibration mode frequencies are not harmonics of the fundamental (unless the membrane is loaded by air, as in the case of a timpani drumhead, or by a patch of applied material, as in the case of the tabla and mrdanga of India. Nodal lines now divide the segments that move in opposite directions. In a circular membrane, these nodal lines tend to be diameters and concentric circles.

Now we consider quite a different kind of vibrator: one that has stiffness to bending, such as a bar or rod. It is possible to transmit transverse (bending) waves, longitudinal (stretching) waves, or torsional (twisting) waves on a bar. Transverse vibrations are most useful in musical instruments, so we begin with these.

5.1. Transverse Vibrations of a Bar or Rod

A bar or rod is capable of transverse vibrations in somewhat the same manner as a string. The dependence of the frequency on tension is more complicated than it is in a string, however. In fact, a bar vibrates quite nicely under zero tension, the elastic forces within the bar supplying the necessary restoring force in this case. In this section, we will present only results; for a discussion of the physics of vibrations, see *Physics of Musical Instruments* [1] or a textbook on engineering mechanics.

The speed of transverse waves in a bar is proportional to the square root of frequency, unlike the speed of transverse waves on a string which is independent of frequency. As a result of this *dispersion* (frequency dependence of wave speed), the transverse vibrational mode frequencies of a bar or rod are not harmonic as they are in a string. The actual mode frequencies depend upon the end conditions.

In books on elasticity, three possible end conditions are generally considered: free ends, clamped ends, and simply-supported (hinged) ends. Of the 9 possible combinations of end conditions, the ones encountered in musical instruments are free-free and free-clamped. The transverse mode frequencies of a bar with rectangular cross section that is free at both ends (free-free) are given by: $f_n = 0.113 h/L^2 \sqrt{E/\rho}\ (2n+1)^2$. In this formula, h is the thickness of the bar, L is its length, E is its elastic modulus, and ρ is the density of its material. Note that the mode frequencies are proportional to the thickness and inversely proportional to the length squared, but independent of the width.

For precise calculations, it should be noted that for the fundamental mode ($n=1$), the factor *2n+1* should be replaced by 3.011, so the frequencies are proportional to $(3.011)^2$, 5^2, 7^2, etc. This is obviously not a harmonic series; the frequencies can be written as f_1, $2.756f_1$, $5.404f_1$, etc. Mode frequencies, wavelengths, and nodal positions for the first four transverse modes in a bar of rectangular cross section with free ends are given in Table 5.1.

Table 5.1. Mode frequencies and node positions in a bar of uniform rectangular cross section with free ends

Frequency (Hz)	Wavelength (m)	Nodal positions (m from end of 1-m bar)
$f_1 = 3.5607 \ /L^2\sqrt{E/\rho}$	1.330L	0.224, 0.776
$2.756f_1$	0.800L	0.132, 0.500, 0.868
$5.404f_1$	0.572L	0.094, 0.356, 0.644, 0.906
$8.933f_1$	0.445L	0.073, 0.277, 0.500, 0.723, 0.927

The transverse mode frequencies of a rod or tube with free ends are similar to those in the bar of rectangular cross section. For a rod of radius *a*, replace $0.113h$ in the f_n formula by $0.196a$, and for a tube with outer radius *a* and inner radius *b*, replace $0.113h$ by $0.196\sqrt{a^2+b^2}$. The relative frequencies of the modes, the dependence on length, and the locations of the nodes stay the same.

A cantilevered bar or rod has one clamped end and one free end. The mode frequencies are given by: $f_n=0.113h/L^2\sqrt{E/\rho}\,(2n-1)^2$. For precision, the factor *(2n-1)* should be replaced by 1.194 and 2.988 in the lowest two modes. The first four frequencies will be in the ratios 1 : 6.267: 17.55: 35.39. The frequency of the lowest mode is only about 1/6 of the corresponding frequency in a bar with two free ends. Bending modes of free-free and free-clamped bars or rods are shown in Fig. 5.1.

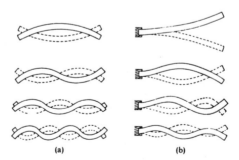

Fig. 5.1. Bending modes of (a) a bar with two free ends (as in a marimba); (b) a bar with one free and one clamped end (as in lamellophones).

5.2. Longitudinal Vibrations of a Bar or Rod

Sound waves traveling in air are longitudinal waves; the air molecules vibrate back and forth in the same direction the wave is propagating. Similar longitudinal (compressional) waves can propagate in solids and liquids, resulting in longitudinal standing waves or normal modes of vibration. In a bar or rod, longitudinal waves travel at a speed $v=\sqrt{E/\rho}$, so the mode frequencies are: $f_n = \frac{1}{2}n/L\sqrt{E/\rho}$. Note that the wave speed does not depend on frequency (no dispersion) and that the expression for the mode frequencies is simpler than in the case of transverse vibrations. The mode frequencies are independent of the thickness or radius of the bar or rod, and they are inversely proportional to the length L (rather than to L^2).

In a bar or rod with free ends, the fundamental mode will have a node at its center. The next mode has two nodes at $\frac{1}{4}L$ and $\frac{3}{4}L$, the third mode has 3 nodes, and so on. Note that only the odd-numbered modes have a node at the center. In most bars or rods, the longitudinal modes of vibration occur at much higher frequencies than the transverse modes.

5.3. Torsional Vibrations of a Bar or Rod

Torsional or twisting waves are a third type of wave motion possible in a bar or rod. The elastic (Young's) modulus E is replaced by the shear modulus G, so the speed of torsional waves in a circular rod becomes $c_T = \sqrt{G/\rho}$. In bars of rectangular cross section, the speed of transverse and longitudinal waves was nearly independent of the width of the bar. Not so with torsional waves. Torsional wave velocities in rods and bars with different cross sections are given in Fig. 5.2.

$c_T = \sqrt{G/\rho}$ $c_T = 0.92\sqrt{G/\rho}$ $c_T = 0.74\sqrt{G/\rho}$ $c_T = (2h/w)\sqrt{G/\rho}$

Fig. 5.2. Torsional wave velocities in rods and bars with different cross sections.

5.4. Vibrations of Air Columns

Although air columns are mainly associated with wind instruments, they are used as resonators in marimbas, xylophones, vibes, and certain drums, so their modes of vibrations will be briefly considered.

Sound waves inside an air column generally propagate at the same speed as waves in free space: $v = 20.1\sqrt{T}$, where T is the temperature on the absolute scale (obtained by adding 273 to the Celsius temperature). At 20°C, for example, $T=293$, so $v=343$ m/s. Over the range of temperature we normally encounter, the speed of sound increases by about 0.6 m/s for each Celsius degrees and an approximate formula for the speed of sound is sufficiently accurate: $v=332.4 + 0.6\,t$, where t is temperature on the Celsius scale. We could also write the speed of sound in feet/second as: $v=1052 + 1.09t'$, where t' is the temperature on the Fahrenheit scale. At 70° F, the speed of sound is 1128 ft/s.

Modes of vibration or resonances of a pipe open at both ends are shown in Fig. 5.3(a). Note that the sound pressure p drops to zero (i.e., the total pressure equals atmospheric pressure) at the open ends, while the net displacement of air (represented by the arrows) is a maximum at the open ends. Minimum displacement occurs at the nodes denoted by N. The fact that the pressure is maximum where the air displacement is minimum (and vice versa) may be a little confusing, at first glance, and this has led to misunderstanding about pipe resonators in various books and papers. Note that the frequencies of the modes in Fig. 5.3(a) are harmonics of the fundamental frequency $v/2L$.

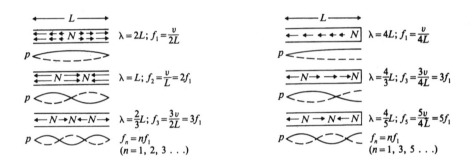

Fig. 5.3. (a) Modes of vibration or resonances of an open pipe. At the open ends the sound pressure is minimum while the air displacement (shown by arrows) is maximum. Minimum air displacement occurs at the nodes denoted by N. (b) Modes of vibration or resonances in a pipe closed at one end. At the closed end, the air displacement is minimum (a node N occurs) but the pressure is maximum. The resulting modes include odd-numbered harmonics only.

Modes of vibration or resonances of a closed pipe (i.e., one end closed, one end open) are shown in Fig. 5.3(b). Note that the sound pressure p is maximum at the closed end but drops to zero (i.e., the total pressure equals atmospheric pressure) at the open end. This leads to an asymmetrical pattern of pressure, so that the lowest mode occurs at a frequency where the pipe length L equals one-fourth of a wavelength. Thus, the fundamental frequency of a closed pipe is one-half that of an open pipe of the same length, a fact that is certainly well known to organ builders. Note that only the odd-numbered harmonics of this fundamental frequency are resonances of a closed pipe.

Marimbas, xylophones, and vibes generally employ closed tubes as resonators, since the tubes only need to be half as long as open tubes would be. This means that only odd-numbered harmonics are supported (fundamental and third harmonic in xylophones but only the fundamental in marimbas and vibes, as we will see in Chapter 6).

5.5. End Correction

The length L in the formulas in the previous section should be understood to be the acoustic length of the pipe, not its physical length. In a pipe with two closed ends (not a very useful resonator), there would be no difference between the acoustic and physical lengths, but at an open end the sound pressure does not drop abruptly to zero. Instead, p drops to zero over a distance that depends upon the diameter of the pipe and its environment, and we add an appropriate end correction to the physical length to obtain the acoustic length. For a circular pipe with no flange, the appropriate end correction at low frequencies is about 0.3 d, where d is the diameter. For a pipe terminating in a large flange, the end correction is about 0.4 d.

When the pipe faces a bar, the end correction increases slightly, since a channel is created between the pipe and bar and the air must move faster to move the same volume. This should be taken into account when designing tube resonators for marimbas or xylophones, for example. The pipe should be a bit shorter in place in the instrument than it would be to resonate at the same frequency in free space. Thus, resonator tubes are tuned by sliding the end cap when the pipe is in playing position. Some instruments provide for temperature compensation by changing the bar-to-pipe distance for different temperatures. At higher temperature the pipe is moved slightly closer to the bar to compensate for the higher sound velocity.

References

1. N. H. Fletcher and T. D. Rossing, *The Physics of Musical Instruments,* 2nd ed. (Springer-Verlag, New York, 1998).
2. T. D. Rossing, *The Science of Sound,* 2nd ed. (Addison-Wesley, Reading, MA, 1990).

Chapter 6
Xylophones and Marimbas

Xylophones and marimbas are both described by musicologists as idiophones with wooden bars. *Xylo-*, in fact, means wood, and in some European countries the term xylophone includes both xylophones and marimbas. Idiophones (*idios* = self) are instruments built around bars or plates or other structures capable, by themselves, of vibrating and producing sound, as compared to drums (membranophones) that require a tensioning mechanism in order to vibrate.

6.1. Xylophones

The xylophone, which has been used in Africa and Asia for many centuries, is one of the oldest melodic instruments known. Early versions consisted of roughly hewn wooden bars held in the player's lap or placed over a hollow log, which may have served as a sort of resonator. Later, instruments with individual gourd resonators were developed.

The modern xylophone has between 3 and 4 octaves of wood or synthetic bars laid out in piano-keyboard fashion, with the sharps and flats in a raised row behind the naturals. The bars are undercut at the center to tune the second vibrational mode to three times the frequency of the fundamental (as compared to 2.76 times the fundamental in a bar of rectangular cross section and uniform thickness, see Section 5.1). A closed-pipe resonator is generally placed under each bar to amplify both the fundamental and the first overtone. The bars sound one octave higher than the written music.

The bars are generally suspended from two cords that pass through holes drilled close to the location of the two nodes in the lowest mode of vibration (see Table 5.1) in order to minimize damping. The use of hard mallets, the tuning of the first overtone to the third harmonic, and the support of the third harmonic by the closed tube resonator gives the xylophone a crisp bright sound. In Saint-Saëns' *Danse Macabre*, for example, it is supposed to depict the rattle of the bones of the dead.

The xylomarimba or xylorimba is a large xylophone with a 4½ to 5-octave range (C_3 or F_3 to C_8), and is occasionally used in solo work or in modern scores. Bass xylophones and keyboard xylophones have also been constructed [1].

Vibrational Modes

The vibrational mode frequencies of a xylophone bar tuned to $F_4^{\#}$ (f=370 Hz) are shown in Fig. 6.1. The frequencies of the bending and torsional modes have been lowered by cutting the arch on the underside of the bar, and this lowering is greatest in the lowest modes in each family. Thus, the slopes of curves connecting the lowest torsional and bending modes in Fig. 6.1 have greater slopes than those for a rectangular bar, although they approach the latter slopes (1 and 2, respectively) with increasing mode number.

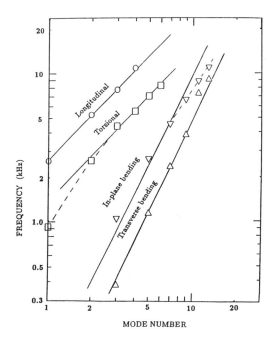

Fig. 6.1. Modal frequencies of an $F_4^\#$ xylophone bar. Solid lines have slopes of 1 or 2. The mode number is $2n+1$ for bending modes (see Section 5.1) [2].

Decay Times

Studies of decay times in wooden xylophone bars without resonators have shown that the decay process is mainly determined by internal losses; radiation losses and friction at the cord support appear to be insignificant. The frequency dependence of the damping constant α or the decay time t_d can be expressed by $\alpha(f) = 1/t_d = a_o + a_1 f^2$, where a_o and a_1 depend on the wood species [3].

Holz [4] compares the acoustically important properties of xylophone bar materials and concludes that the "ideal" wood is characterized by a density of 0.80 to 0.95 g/cm³ (800 to 950 kg/m³) and an elastic (Young's) modulus of 15 to 20 GPa. These conditions are met by several Palissandre species, some other tropical woods, and also by cherry wood. In order to be suitable substitutes for wood, glass-fiber reinforced plastics should be pressed and molded, rather than hand lay-up laminates, Holz cautions.

Elastic properties of some wood and metal materials are given in Table 6.1. It is interesting to note that the sound velocity (which depends on $\sqrt{E/\rho}$, see Section 5.1) is quite similar in aluminum, steel, glass, African mahogany, redwood, and Sitka spruce.

Table 6.1. Elastic properties of materials [5]

Material	Density $\rho(\text{kg/m}^3)$	Young's modulus $E(\text{N/m}^2)$		Sound velocity $v(\text{m/s})$		Reference
Aluminum	2700	7.1×10^{10}		5150		Kinsler et al., 1982
Brass	8500	10.4×10^{10}		3500		
Copper	8900	10.4×10^{10}		3700		
Steel	7700	19.5×10^{10}		5050		
Glass	2300	6.2×10^{10}		5200		
Wood		∥ grain	⊥ grain	∥ grain	⊥ grain	
Brazilian rosewood	830	1.6×10^{10}	2.8×10^9	4400	1800	Haines, 1979
Indian rosewood	740	1.2×10^{10}	1.7×10^9	4000	1500	
African mahogany	550	1.2×10^{10}	1.2×10^9	5000	1500	
European maple	640	1.0×10^{10}	2.2×10^9	4000	1800	
Redwood	380	0.95×10^{10}	0.96×10^9	5000	1600	
Sitka spruce	470	1.3×10^{10}	1.3×10^9	5200	1700	

6.2. Marimbas

The term marimba has different meanings in different musical cultures. In eastern and southern Africa it denotes a group of idiophones, some of which are struck and some of which are plucked [6]. In Latin America it is mostly used for the calabash-resonated xylophone introduced from Africa, although in Colombia it is used to denote any melodic instrument other than wind instruments. The term is now used in most parts of the world to denote the deep-toned orchestral marimba with tuned bars and resonator tubes that evolved from the early Latin American instrument.

Much credit for developing the modern orchestral marimba is generally given to J. C. Deagan, an English-born clarinetist who emigrated to the United States in 1879 and began making percussion instruments in Chicago. The instrument grew in popularity, partly through the efforts of Clair Musser, who organized large-scale marimba ensembles and gave a concert in Carnegie Hall in 1935 with his 100-piece marimba band (see Section 6.5).

The marimba typically includes 3 to 4½ octaves of tuned bars of rosewood or synthetic material, graduated in width from about 4.5 to 6.4 cm (1¾ to 2½ in). Beneath each bar is a closed pipe resonator tuned to the fundamental frequency of that bar. When the marimba is played with soft mallets, it produces a rich mellow tone. The playing range of a large concert marimba is typically A_2 (110 Hz) to C_7 (110 to 2093 Hz), although bass marimbas extend down to C_2 (65 Hz). A few instruments cover a full five-octave range.

A deep arch is cut in the underside of marimba bars, particularly in the low register. This arch serves two useful purposes: it reduces the length of the bar required to reach the low pitches, and it allows tuning of the overtones (the first overtone is nominally tuned two octaves above the fundamental). Fig. 6.2 shows a scale drawing of a marimba bar and also indicates the positions of the nodes for each of the first seven modes of vibration. Note that

Marimbas and Xylophones

the number of nodes increases by one in each successive mode. The even-numbered modes have a node at the center, while the odd-numbered modes have an anti-node (maximum amplitude) there.

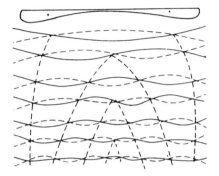

Fig. 6.2. Scale drawing of a marimba bar tuned to E_3 (165 Hz). The dashed lines locate the nodes of the first seven modes [7].

The sound spectrum of the same E_3 marimba bar, shown in Fig. 6.3 along with the partials on musical staves, indicates the presence of a strong third partial with a frequency of 9.2 times the fundamental (about three octaves plus a minor third above it). The relative strengths of the partials, of course, depend on where the bar is struck and what type of mallet is used. To emphasize a particular partial, the bar should be struck at a point of maximum amplitude for that mode, as indicated in Fig. 6.2.

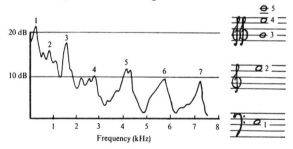

Fig. 6.3. Sound spectrum for an E_3 marimba bar. The partials are also indicated on music staves, which include a "super-treble" clef, two octaves above the treble clef [7].

Figure 6.4 compares the vibrational behavior of a marimba bar and a uniform rectangular bar for the first five bending modes. The upper row shows the local bending force, and the second row the bending moment. The slope of the bar is shown in the third row and the displacement in the fourth row. Note that the maximum displacement for all modes of the uniform bar occurs at the ends, but in the marimba bar it occurs near the center.

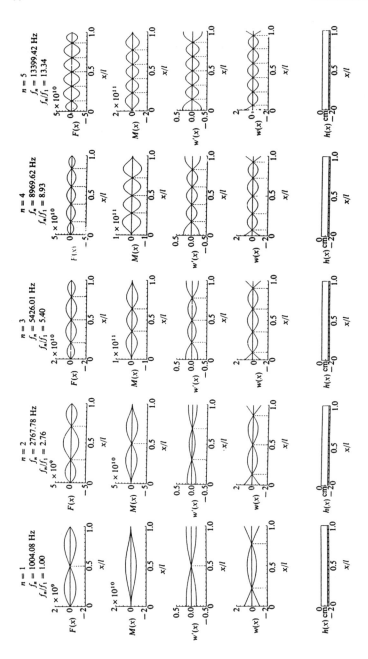

Fig. 6.4 (a) Spatial distribution of force, bending moment, and displacement for the first five modes of a uniform rectangular bar [8].

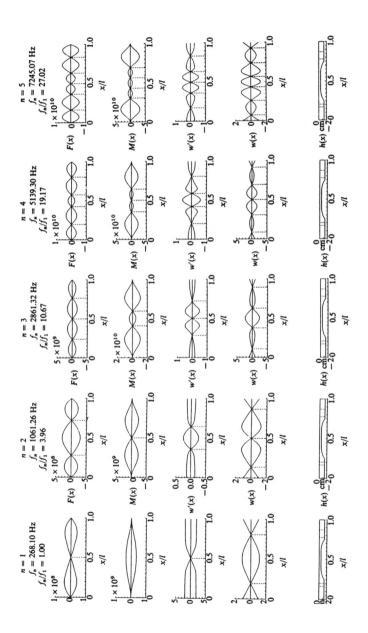

Fig. 6.4 (b) Spatial distribution of force, bending moment, and displacement for the first five modes of a marimba bar [8].

Oscillograms of the first three partials from an A_3 (220 Hz) marimba bar are shown in Fig. 6.5. Note that the second and third partials appear and also disappear much more quickly than the fundamental. In fact, the third partial has completely decayed before the fundamental has reached its maximum amplitude. It has been found that the tuning of the inharmonic third partial has some slight influence on the perceived pitch of the bar [9].

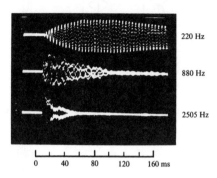

Fig. 6.5. Oscillograms of the first three partials from an A_3 marimba bar. The second and third partials have been amplified for clarity [9].

6.3. Tuning the Bars

Removing material from any point on a bar affects all the modal frequencies to some extent. However from Fig. 6.4 it is clear that removal from certain places affects some modes more than others, and thus it is possible to tune the individual partials. In general, removing material from a place where the bending moment M(x) for a given mode is large will lower the frequency of that mode considerably. It is more difficult to raise the frequency of a given mode, but removing material near the end of the bar will slightly raise the frequency of all the modes (even though the length remains unchanged).

Bork [8] has made a careful study of bar tuning. Figure 6.6 shows the effect of a small lateral cut at various positions on the first four modes of a rectangular bar and a bar with the center portion thinned so that $f_2 = 4f_1$. The degree which each partial is raised or lowered is readily seen. Note that the second mode of the marimba bar (Fig. 6.6b) is easily tuned by removing material at about one-third the distance from each end, whereas the lowest mode is most affected by thinning the center.

Bork and Meyer [9] investigated the tuning of the third partial to various ratios, both harmonic and inharmonic with respect to the fundamental. They compared synthesized bar sounds in which the third partial was tuned to the following six musical intervals above the third octave: major second, major second +50 cents, minor third, minor third +50 cents, major third, and major third +50 cents. They found a preference for the fourth choice, where the third partial is tuned midway between a major and minor third above the triple octave (i.e., $f_3 = 9.88 f_1$). (Tuning it midway between a major third and a fourth also proved satisfactory; in this case a brighter timbre resulted).

Table 6.2 gives numerical calculations by Orduña-Bustamante [11] for parabolic undercuts resulting in 7 harmonic tunings with f_2/f_1 ranging from 3 to 5 and f_3/f_1 ranging from 6 to 13. R_{21} and R_{31} are frequency ratios, X_c and T_c are dimensionless measures of the width of the undercut and the minimum thickness of the bar.

Table 6.2. Frequency ratios and parameters of the parabolic undercut of free rectangular bars with harmonic overtones [10]

Beam	R_{21}	R_{31}	X_c	T_c
1	3	6	0.4163	0.7340
2	4	8	0.1642	0.5073
3	4	9	0.2478	0.4518
4	5	10	0.1282	0.3722
5	5	11	0.1595	0.3590
6	5	12	0.1898	0.3361
7	5	13	0.2236	0.3020

While the parabolic contours in Table 6.2 can easily be cut with a numerically controlled milling machine, in practice fine tuning is necessary, especially with bars of natural wood. Orduña-Bustamante gives curves showing contours of equal frequency that can be used as a guide to fine tuning the bars [10]. Figure 6.6 shows calculated parabolic contours and final contours for bars having harmonic ratios 1:3:6, 1:4:8, and 1:4:9, respectively.

Summers, et al. show that f_2/f_1 ratios of 3 or 4 can be obtained with a simple rectangular cut. Discussions of practical marimba tuning can be found in MacCallum [11], Moore [12], and Bork [13].

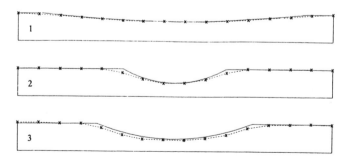

Fig. 6.6. Contours of bars calculated mathematically (solid lines) and bars tuned for exact harmonic ratios (dashed lines). Beam 1) 1:3:6; Beam 2) 1:4:8; Beam 3) 1:4:9 [10].

6.4. Resonators

Marimba resonators are cylindrical pipes tuned to the fundamental mode of the corresponding bars. A pipe with one closed end and one open end resonates when its acoustical length is one-fourth of a wavelength of the sound. The purpose of the tubular resonators is to emphasize the fundamental and also to increase the loudness, which is done at the expense of shortening the decay time of the sound. The statement is sometimes made that the resonator prolongs the sound, but this is incorrect. In a noisy environment or when played with other instruments in an ensemble, that impression may be conveyed, as can be understood by referring to Fig. 6.7.

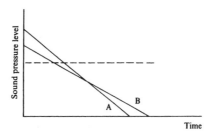

Fig. 6.7. Sound decay for a marimba bar with a resonator (curve A) and a bar without a resonator (curve B). Note that curve A begins at a greater sound level but the decay time is shorter. If the background level (dashed line) is high enough, the sound represented by curve A may appear to the longer in duration.

Curve A, which represents the more rapid decay of a bar with a resonator, begins at a higher sound level than curve B, which represents the decay of a bar with no resonator. At some time, the curves cross. If the level of background noise (dashed line in Fig. 6.7) is high enough, curve A may cross it after curve B, thus appearing to be longer in duration even though the decay time is shorter.

What the resonator does is to increase the radiation efficiency of the system. Without a resonator, the bar radiates essentially as a dipole source (see Section 2.6) with considerable flow of air back and forth from the top side to bottom side of the bar. The resonator upsets the balance between the out-of-phase sources, introducing a more efficient monopole component into the radiating system.

Decay times are often expressed as the time that would be required for the sound level to fall 60 dB (even though it fades into the background noise before then). The 60-dB decay time of a typical rosewood marimba bar in the low register (E_3) is about 1.5 s with the resonator and 3.2 s without it. Decay times in the upper register are generally shorter; we measured 0.4 s and 0.5 s for an E_6 bar with and without the resonator, respectively. The corresponding decay times for synthetic bars are slightly longer.

Pipes that are closed at one end and open at the other have resonances that are nearly the odd-numbered harmonics of the fundamental mode (i.e., the resonance frequencies would be f, $3f$, $5f$, ...). Thus, a resonator tuned to the fundamental frequency of a marimba bar would not affect the first overtone, because the frequency of that mode is four times the fundamental frequency. As an experiment, a marimba having a second set of resonators tuned to the first overtone of the bars was constructed. Each resonator was equipped with a vane that can partially or completely close the mouth of the tube; thus the timbre can be varied by adjusting the amount of closure [14].

The resonance frequency of a pipe depends, to a small extent, on the environment near the open end. This is because the reflection of standing sound waves within the tube does not take place precisely at the end of the tube, as discussed in Section 5.5. If the bar is close to the resonator tube, the resonance frequency will be affected by its presence. In particular, the resonance frequency will be lowered as the bar is moved closer to the tube. This effect can be used to compensate for changes in resonator frequency caused by changes in ambient temperature. It works as follows: if the temperature rises, the speed of sound increases, and so does the resonance frequency of the tube. Moving the resonator closer to the bar, however, lowers the resonance frequency of the tube, so that it once again matches that of the bar. Some marimbas incorporate an adjustment of this nature.

Resonators with two open ends, having twice the length of conventional resonators, offer an interesting alternative at the higher frequencies. Because both ends of the bar radiate sound, both constructive and destructive interference occur, and the sound field becomes more directional. Fig. 6.8 compares the sound pressure levels at three measuring points with a conventional closed end quarter-wave ($\lambda/4$) resonator and a half-wave ($\lambda/2$) resonator with both ends open. Owing to constructive inference, the $\lambda/2$-resonator increases the sound field by approximately 3 dB at MP2, which is in the far field in the mid-plane of the resonator. However, the sound level at the position of the performer (MP3) is lowered considerably. Relatively little difference between the $\lambda/4$ and $\lambda/2$ resonators occurs at MP1 in the plane of the bar [8].

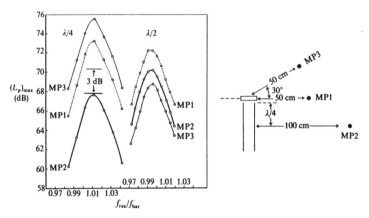

Fig. 6.8. Sound pressure levels as functions of frequency at three different points for $\lambda/4$ (one closed end) and $\lambda/2$ (both ends open) resonators [8].

Neighboring resonators also have an effect on sound radiation. A nearby resonator tuned a semitone below the bar can decrease the sound output by as much as 2 dB, whereas a nearby resonator tuned a semitone higher can raise the sound output by about 1 to 1.5 dB [8]. This interaction would be most noticeable, therefore, at the semitone intervals E to F and B to C. A similar interaction becomes important in bass marimbas with trough resonators.

6.5. Marimba Orchestras and Clair Musser

Clair Omar Musser distinguished himself as a marimba virtuoso, designer, composer, and arranger. It was probably as organizer and conductor of large marimba orchestras that he most distinguished himself, however. He organized a 25-marimba "all girl" orchestra for Paramount Pictures and their opening performance at the Oriental Theatre in Chicago in 1929. He designed a special marimba for a 100-piece orchestra that he trained and directed for the Century of Progress Exposition in Chicago in 1933. Then he designed the King George Marimba for the famous 100-piece Imperial Marimba Symphony Orchestra of 50 women and 50 men that toured Europe.

His marimba orchestras got larger and larger. In 1941 Musser conducted a 150-piece orchestra in Chicago, and in 1949 he conducted a 200-piece orchestra sponsored by the Chicago Tribune at Soldier's Field in a concert heard by 111,000 people. His largest marimba orchestra was a 300-piece orchestra that performed on the North State at the Chicago Fair September 1, 1950 along with a 100-voice choir and a battery of contra-bass marimbas.[15]

6.6. Mallets

The modern percussion player can select from a wide variety of mallets that differ in mass, shape, and hardness. Through intelligent selection of a mallet, a player can greatly influence the timbre of the instrument being played. Striking a marimba or xylophone with a hard mallet, for example, produces a sound rich in overtones that emphasizes the woody character of the instrument. A soft mallet, on the other hand, which excites mainly the harmonically tuned lower partials, gives a dark sound to the instrument.

A mallet whose mass nearly equals the dynamic mass of the struck bar (typically about 30% of the total mass for a marimba bar in its fundamental mode) transfers the maximum amount of energy to the bar. A lighter mallet rebounds after a short contact time. A heavier mallet remains in contact for a longer time, which results in considerable damping of the higher partials. This is a desirable effect in drums, where a large amplitude and short decay time are desired, but may be only a special effect in bar percussion instruments.

According to Hertz's law, the impact force is proportional to the 3/2 power of the mallet deformation. From measurements of impact force and impact time, it appears that Hertz's law describes the impact over a fairly wide range of mallet hardness and impact time [3].

References

1. R. S. Brindle, *Contemporary Percussion.* (Oxford Univ. Press, London, 1970).
2. T. D. Rossing and D. A. Russell, *American J. Physics* **58** (1990) 1153.
3. A. Chaigne and V. Doutaut, *J. Acoust. Soc. Am.* **101** (1997) 539.
4. D. Holz, *Acustica* **82** (1996) 878.
5. N. H. Fletcher and T. D. Rossing, *The Physics of Musical Instruments* (Springer-Verlag, New York, 1998) Chapter 19.
6. J. Blades, *New Grove Dictionary of Musical Instruments*, vol. 2, ed. S. Sadie (Macmillan, London, 1984) 617.
7. T. D. Rossing, *Physics Teacher* **14** (1976) 546.
8. I.Bork, *Zur Abstimmung and Kopplung von Schwingen den Stäben und Hohlraumresonatoren* (Dissertation, tech. Univ. Carolo-Wilhelmina, Braunschweig, 1983).
9. I. Bork and J. Meyer, *Das Musikinstrument* **31**(8) (1982) 1076. (English translation in *Percussive Notes* **23**(6) (1985) 48).
10. F. Orduña-Bustamante, *J. Acoust. Soc. Am.* **90** (1991) 2935.
11. F. K. MacCallum, *The Book of Marimba* (Carlton Press, New York, 1968).
12. J. Moore, *Acoustics of Bar Percussion Instruments* (Ph.D. thesis, Ohio State University, Columbus, 1970).
13. I. Bork, *Applied Acoustics* **46** (1995) 103.
14. G. Kvistad and T. D. Rossing, *J. Acoust. Soc. Am.* **61** (1977) S21.
15. Anon., *Percussive Notes* **37**(2), 6-9 (1999).

Chapter 7
Metallophones

In this chapter, we will consider idiophones with vibrating bars and rods of steel, aluminum, and bronze. Some of these instruments are struck with hard mallets to give a very metallic sound, others are struck with soft mallets to give a more mellow sound, some are plucked. Some have resonators, some do not.

In addition to the familiar metallophones played in Western orchestras and ensembles, many different types of metallophones are used in other cultures, such as Asian and African. We will discuss a few of these ethnic metallophones as well as some contemporary instruments based on aluminum tubes.

7.1. Orchestra Bells or Glockenspiel

The glockenspiel has 2½ octaves of steel bars with rectangular cross section arranged in the manner of a piano keyboard. The range is generally G_5 to C_8, scored two octaves lower than sounded. The bars, about 6 to 9 mm (¼ to ⅜ in) thick and 25 to 32 mm (1 to 1¼ in) wide, are generally mounted in a portable case with a removable lid, and the rear row of bars ("black keys") are raised above the front bars, as shown in Fig. 7.1(a). Because the overtones occur at such high frequencies, where the ear is not very sensitive to pitch, the upper partials of the bar are not tuned by undercutting.

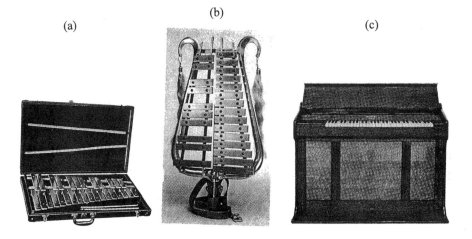

Fig. 7.1 (a) Glockenspiel; (b) bell lyra; (c) celesta.

The vibrational modes of a glockenspiel bar are shown in Fig. 7.2. Besides the transverse modes, labeled 1, 2, 3, 4, and 5, there are torsional or twisting modes labeled *a, b, c,* and *d*; a longitudinal mode *l*; and transverse modes in the plane of the bar 1*x* and 2*x*.

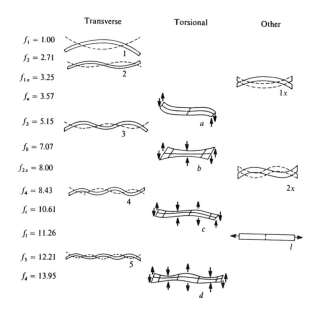

Fig. 7.2. Vibrational modes of a glockenspiel bar, with relative frequencies at the left [1].

The frequencies for transverse modes of vibration in a bar with free ends were shown to be inharmonic (see Section 5.1). The transverse modes of vibration of the glockenspiel bar are spaced a little closer together than predicted by the simple theory of thin bars, discussed in Section 5.1, but they are still quite inharmonic. Note that the four torsional modes in Fig. 7.2. have frequencies in the ratios 2.02 : 2.00 : 3.00 : 3.94, however, which makes them nearly harmonic. Similarly, the frequencies of the longitudinal modes (of which only the lowest mode is shown in Fig. 7.2) form a harmonic series, as shown in Section 5.2. This is of little musical importance in the glockenspiel, however, since these modes are not excited in normal playing.

The glockenspiel is played with a variety of beaters: ebonite, wood, plastic, and brass for a loud, bright sound and mallets with soft rubber heads for soft passages. The glockenspiel often doubles the melody played by another instrument, adding sparkle and clarity to the sound. As with other mallet percussion instruments, glissando and tremolo can be used in forte passages. Chords are infrequently played.

The *bell lyra* (Fig. 1(b)) is a portable form of glockenspiel adapted for use in marching bands. The keyboard is held vertical in a lyre-shaped aluminum frame fixed to a staff which is carried in the player's belt. The instrument can be fixed vertically to a stand when not on the march. Bell lyras have about a 2-octave range, slightly smaller than that of a glockenspiel, and are generally played with hard plastic mallets.

7.2. Celesta

The celesta, shown in Fig. 1(c), could be described as a glockenspiel played with felt hammers by means of a piano-type keyboard. Each steel bar has its own box resonator, and there is a sustaining pedal. The range is 4 to 5 octaves, and the music is written an octave lower than it sounds. In orchestras, the celesta is generally played by the keyboard players rather than the percussionists.

7.3. Vibraphone or Vibes

The vibraphone or vibraharp, commonly called vibes, has tuned aluminum bars, covering about a three-octave range from F_3 to F_6. The bars are deeply arched so that the first overtone is two octaves above the fundamental (i.e., it has four times the fundamental frequency). The aluminum bars tend to have much longer decay times than the wood or synthetic bars of the marimba or xylophone, and so vibes are equipped with pedal-operated dampers. They are usually played with soft yarn-covered mallets, which produce a mellow tone, although some passages call for harder beaters.

The most distinctive feature of the vibes is the vibrato introduced by motor-driven disks at the tops of the resonators, which alternately open and close the tubes. The vibrato produced by these rotating disks consists of a rather substantial fluctuation in amplitude (intensity vibrato) combined with a small change in frequency (pitch vibrato). The speed of rotation of the disks may be adjusted to produce a slow vibe or a fast vibe. Sometimes vibes are played without vibrato by switching off the motor driving the disks.

Because vibraphone bars have a much longer decay time than do marimba and xylophone bars, the effect of the tubular resonators on decay time is quite dramatic. At 220 Hz (A_3), for example, the decay time (60 dB) is 40 s without the resonator and only 9 s with the tube fully open. For A_5 the decay time is 24 s with the resonator closed and 8 s with it open [1]. In the sound level recording in Fig. 7.3, the intensity modulation, as well as the slow decay of the sound can be clearly seen.

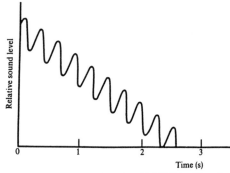

Fig. 7.3. Sound level recording for an A_3 vibraphone bar (f=440 Hz) with a vibe rate of about 4 Hz. The level fluctuates about 6 dB, and the decay time (60 dB) is 7 s [2].

The mode frequencies for three vibe bars tuned to F_3, G_4, and G_5 are given in Table 7.1. Note that while the second mode ($n=2$) is tuned to about four times the fundamental frequency, the frequencies of the higher modes change quite markedly in relation to the fundamental. For G_5 and higher, the second partial is only approximately tuned to the double octave. The same is true in most marimbas and xylophones.

Table 7.1. Mode frequencies of three vibe bars along with ratios to the fundamental [2].

n	f_n	f_n/f_1	f_n	f_n/f_1	f_n	f_n/f_1
1	175	1	394	1	784	1
2	700	4.0	1578	4.0	2994	3.8
3	1708	9.7	3480	8.9	5995	7.6
4	3192	18.3	5972	15.2	9400	12.0
5	4105	23.5	8007	20.2	14,014	17.9
6	6173	35.4	11,119	35	18,796	24.0
7	8080	46.3			21,302	27.2

7.4. Interlude: Thick Bars vs Thin Bars

Transverse vibrations of thin bars and rods were discussed in Section 5.1. The restoring force in a thin bar or rod is assumed to be due to the bending moment only. Such a simplified model leads to what is generally called Euler-Bernoulli beam theory. The Timoshenko beam theory, which considers rotary inertia and shear stress, is preferred for thick bars and rods.

As a beam bends, the various elements rotate through some small angle. The rotary inertia is thus equivalent to an increase in mass and results in a slight lowering of vibrational frequencies, especially the higher ones.

Shear forces, which are present in a bar or rod when it is bent, tend to deform the bar; in particular they cause rectangular elements to become parallelograms and thus to decrease the transverse deflection slightly. Therefore, the frequencies of the higher modes are decreased slightly in a thick bar as compared with a thin one.

Thus, we see that both rotary inertia and shear forces tend to decrease the frequencies of the higher bending modes in a thick bar or rod below the values predicted by the simpler thin beam theory.

7.5. Chimes or Tubular Bells

Chimes or tubular bells are widely used in orchestras and bands to produce a sound with a bell-like quality. They are generally fabricated from lengths of brass tubing 32 to 38 mm (1¼ to 1½ in) in diameter. The upper end of each tube is partially or completely closed by a brass plug with a protruding rim which forms a convenient and durable striking point.

Frequencies of the bending modes of a G_4 chime as a function of mode number m are shown in Fig. 7.4 (the mode number m is one greater than the number of nodes). Also shown are the frequencies predicted by a thin-beam theory (Rayleigh) and thick-beam theory

(Flügge). The mode frequencies for the same chime with the end plug removed are also shown in Fig. 7.4. Note that the end plug lowers the frequencies of the lowest modes but has little effect on the higher ones. The lowest five mode frequencies are represented on musical staves in Fig. 7.4, along with the strike note that is heard when the chime is played.

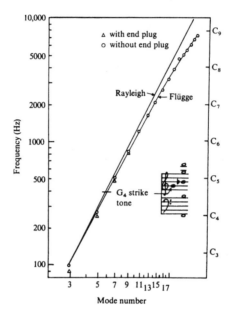

Fig. 7.4. Frequencies of a G_4 chime as a function of mode number m along with those predicted by thin-beam theory (Rayleigh) and thick-beam theory (Flügge). The first five modes are shown on musical staves, along with the subjective strike tone [2].

One of the interesting characteristics of chimes is that there is no mode of vibration with a frequency at, or even near, the pitch of the strike note one hears. This is an example of virtual pitch, discussed in Section 2.5. Modes 4, 5, and 6 appear to determine the strike tone. This can be understood by noting that these modes for a bar with free ends have frequencies in the ratios $9^2:11^2:13^2$, or 81:121:169, which are close enough to the ratios 2:3:4 for the ear to consider them nearly harmonic and to use them as a basis for establishing a virtual pitch. The largest near-common factor in the numbers 81, 121, and 169 is 41.

The ratios of the modal frequencies of a chime tube with and without a load at one end are given in Table 7.2. Also given are the ratios considered desirable for a tuned carillon bell. Note the similarity between the tuning of partials 3 to 8 of a chime and those of a carillon bell. Adding a load to one end of a chime lowers the frequencies of the lower modes more than the higher ones and thus stretches the modes into a more favorable ratio. The end plug also adds to the durability of the chime, and helps to damp out the very high modes.

Table 7.2. Ratios of mode frequencies for loaded and unloaded chime tube [2]

n	Thin rod	Tube	Loaded with			Tuned bell
			193 g	435 g	666 g	
1	0.22	0.24	0.24	0.23		0.5
2	0.61	0.64	0.63	0.62	0.61	1
3	1.21	1.23	1.22	1.22	1.22	1.2
4	2	2	2	2	2	2
						2.5
5	2.99	2.91	2.93	2.95	2.94	3
6	4.17	3.96	4.01	4.04	4.03	4
7	5.56	5.12	5.21	5.21	5.18	5.33
8	7.14	6.37	6.50	6.43	6.37	6.67

The bell-like quality of chimes can be maximized by selecting the optimum size of end plug for each chime. For the G-chime in Fig. 7.4, the 193-g plug, with which the chime was originally fitted, is pretty near the optimum. Most chime makers use the same size plug throughout the entire set of chimes, however, so the timbre changes up and down the scale, typically being optimum near the center. A set of chimes scaled to have the same timbre throughout the entire range would have some advantages. A further step toward an optimum scaling would vary the tube diameter from larger at the low-frequency end to smaller at the high-frequency end.

A well-tuned chime not only has its overtones tuned to resemble those of a tuned bell but it is also free of beats between modes with nearly, but not quite, the same frequencies of vibration. These beating modes most often occur when the chime is not perfectly round or its wall thickness is not perfectly uniform. As a result, the bending vibrations have slightly different frequencies in two different transverse directions, resulting in beats when both modes are excited. These undesired beats can be eliminated by squeezing the chime in a vise or by thinning the wall slightly on one side to bring the modes into tune.

7.6. Triangles and Pentangles

Triangles are generally steel rods bent into a triangle (usually, but not always, equilateral) with one open corner. They are typically available in 15-cm, 20-cm, and 25-cm (6-in, 8-in, and 10-in) sizes, although other sizes are used as well. Sometimes one end of the rod is bent into a hook, or the ends may be turned down to smaller diameters than the rest of the triangle to alter the modes of vibration. They are suspended by a cord from one of the closed corners and are struck with a steel rod or hard beater. Because of their many modes of vibration, triangles are characterized as having an indefinite pitch.

Some of the modal frequencies observed in a 25-cm (10-in) triangle are shown in Fig. 7.5, along with those calculated for a steel rod of the same length and diameter. Modes of vibration in the plane are slightly higher in frequency than the corresponding modes perpendicular to the plane. The triangle modes show a surprisingly close correspondence to those of a straight rod.

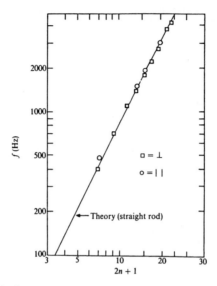

Fig. 7.5. Mode frequencies for a 25-cm steel triangle vibrating in its plane and perpendicular to the plane. The line gives the predicted frequencies for a steel rod of the same diameter and length [2].

The sound of a triangle depends upon the strike point as well as the hardness of the beater. Single strokes are often played on the base of the triangle and perpendicular to the plane [3] to emphasize vibration in the plane of the triangle. A grazing stroke in the upper third of the open leg is recommended for the most even onset and decay of the many partials [4]. Sound spectra for two different strike points with strokes parallel and perpendicular to the plane of the triangle are shown in Fig. 7.6

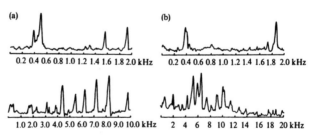

Fig. 7.6. Sound spectra for a 25-cm steel triangle (a) struck in the plane and (b) struck perpendicular to the plane. Two frequency and amplitude ranges are shown in each case [2].

In normal orchestral use, only the higher modes of the triangle are important, since the radiation efficiency of the lower modes is small because of the small diameter of the vibrating rod. However, Australian composer Moya Henderson has developed a new family of instruments based upon triangles or similar bent-rod vibrators coupled to resonators or radiators so arranged as to enhance the lower vibrational modes. These instruments are called alembas.[5]

The first alemba was based on metal triangles coupled by means of light cords to thin diaphragms closing the ends of appropriately tuned pipes. As a part of this development, Dunlop [6] used finite element methods to calculate vibrational frequencies of five in-plane modes (corresponding to $n=4$ to 8 in Fig. 7.5) for triangles with base angles of 30°, 45°, and 60° and various radii of curvature. His calculations indicate that the modes of a triangle would have frequencies somewhat lower than those of a straight rod, although the differences are least, in most cases, when the parameters of an orchestral triangle are approached (60°, small radius of curvature). In a 15-cm (6-in) orchestral triangle, he measured in-plane frequencies for modes with four, five, and seven nodes that have ratios within 5% of those of a straight rod of the same dimensions.

7.7. Gamelan Metallophones

Gamelan is a term used for various orchestras and ensembles in Indonesia and neighboring countries. Most gamelan instruments are percussion instruments, although string and wind instruments and singers may be included. Gamelan can vary in size from a few instruments to seventy-five or more. Gamelan are widely used to accompany dancing and drama, and they are also used in important religious and secular ceremonies. Different gamelan traditions are found in different parts of Indonesia, such as Western Java, Central Java, Eastern Java, and Bali.

Important melodic instruments in every gamelan are keyed metallophones consisting of bronze bars suspended over resonators. In the *gender* type, the metal bars hang over bamboo resonators, while in the *saron* or *gangsa* type, the bars are set on a cushion over a box resonator. Most Metallaphones are played with a small wooden mallet, although the larger gender are played with a padded stick similar to that used in striking the gongs.

A complete gamelan may include three sizes of saron, two sizes of gender, *slentem* (similar to gender but deeper in tone and a single octave only), three sizes of *bonang* (sets of horizontal gong-chimes), *kenong* and *ketuk* (single boxed gongs), at least three sizes of vertically suspended gongs, and various drums (see Section 4.10).

There is no "standard" tuning for gamelan, and it is said that no two gamelan are tuned exactly alike. However, most instruments are tuned either in a five-tone *slendro* or seven-tone *pelog* scales. A typical tuning of Javanese gamelan instruments is illustrated in Fig. 7.7. Intervals between notes are indicated in cents, and a scale of equal temperament is shown for comparison. In the pentatonic slendro scale, intervals tend toward a uniform size, while the intervals in pelog vary markedly in size.

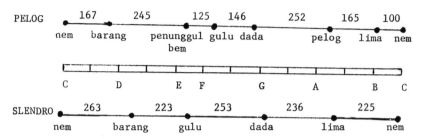

Fig. 7.7. Typical tunings of Javanese gamelan instruments. Numbers of cents in each interval are given, as well as a scale of equal temperament.[7]

A typical gender has 12 to 14 thin bronze keys suspended over individual tube resonators and played with padded disk-shaped mallets, as shown in Fig. 7.8(a). In a complete Central Javanese gamelan, there will typically be three lower-pitched *gender barung* and three higher-pitched *gender panerus,* one tuned in slendro and two in pelog. A much smaller four-bar *gender pemade* from a Balinese gamelan anklung is shown in Fig. 7.8(b).

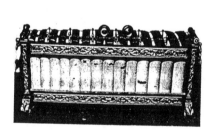 

Fig. 7.8. (a) Gender panerus (Central Java) and (b) gender pemade (Bali).

The sound spectra of the gender pemade in Fig. 7.8(b), at the time of striking and one-half second later, are shown in Fig. 7.9. Note the rapid decay of all but the fundamental. The same is true of most of the gamelan key metallophones we have observed. Frequencies of the first five partials for the bars in a Balinese *jegogan* are given in Table 7.3, along with the average ratios to the fundamental. Since the bars are normally struck near the center, the most prominent overtone is the second one, having a frequency of about 5.2 times the fundamental.

Metallophones

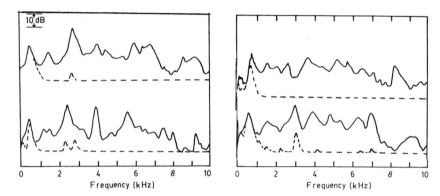

Fig. 7.9. Sound spectra of four bars of a Balinese gender pemade. Solid curves are the spectra at the time of striking, and the dashed curves are the spectra one-half second later. Note the rapid decay of all but the fundamental. [8].

Table 7.3. Vibration frequencies of jegogan bars, along with average frequency ratios [8]

Bar	f_1(Hz)	f_2(Hz)	f_3(Hz)	f_4(Hz)	f_5(Hz)
I	341	903	1796	2953	3998
II	288	771	1497	2444	3519
III	261	694	1348	2199	3186
IV	228	610	1187	1964	2797
Ratio	1	2.7	5.2	8.4	12.2

7.8. Wind Chimes

Wind chimes generally consist of metal tubes or rods suspended so that the blowing wind activates a clapper that strikes all of the tubes or rods in turn. When these tubes are carefully tuned, wind chimes become sonorous musical instruments.

Woodstock Percussion in Shokan, New York began making tuned wind chimes in 1979, and since that time more than 5 million sets of Woodstock Chimes have been sold. Percussionist Garry Kvistad, who founded Woodstock Percussion, selected 5 different pentatonic scales for his Chimes of Olympos (Fig. 7.10a), Chimes of Partch, Chimes of Lun, Chimes of Java, and Chimes of Bali. Their catalog now lists more than 20 different wind chimes, including the Gregorian family of chimes shown in Fig. 7.10(b), consisting of soprano, alto, tenor, and baritone chime sets with eight tubes each, and a "little Gregorian" set with two tubes.

 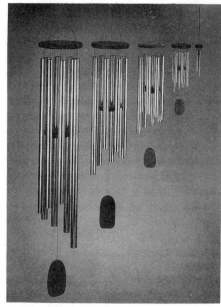

Fig. 7.10. Woodstock chimes. (a) Chimes of Olympos with six aluminum tubes; (b) Gregorian series consisting of soprano, alto, tenor, baritone, and "little Gregorian" chimes.

7.9. Tubaphones, Gamelan Chimes and Other Tubular Metallophones

Metal tubes have been used in various types of keyboard metallophones. The tubaphone, developed in England, consists of brass or steel tubes in a keyboard arrangement similar to a glockenspiel. Although scored in Khachaturian's "Gayané" ballet, the instrument is quite rare in modern orchestras. The Deagan Tubaphone, a version patented by J. C. Deagan in 1889, is shown in Fig. 7.10.

Fig. 7.11. Deagan Tubaphone.

Aluminum tubes are used in a rather simple instrument called Gamelan Chimes, which was developed by percussionist Garry Kvistad especially for home music makers. A set of tubes, which will be used in a particular mode or selection to be played, are placed in rubber cradles and struck with soft mallets. The modes of vibration are similar to those of glockenspiel bars, but the cradles in which the bars rest tend to damp out all but the fundamental mode rather quickly. Tuned aluminum tubes are also used in an instrument called Pipe Dreams, shown in Fig. 7.12.

Fig. 7.12. Pipe Dreams, consists of 2½ octaves of aluminum tubes tuned to a pentatonic scale.

Aluminum tubes, because of their long decay times and their ease of tuning, have been used by several contemporary instrument builders. Lydia Ayers uses more than 200 tubes in her Woodstock Gamelan shown in Fig. 7.13. Her expandable instrument includes several racks of tubes with diameters of 25 to 30 mm (1 to 1.2 in) plus 26 larger hanging tubes similar to orchestra chimes. There are 75 tones in the middle octave, allowing for experimentation in custom tunings [9].

Fig. 7.13. Lydia Ayers and her Woodstock gamelan consisting of more than 200 tuned aluminum tubes.

In his zoomoozophone, Dean Drummond uses 129 aluminum tubes tuned to a 31-note per octave scale. The instruments is divided into 5 sections, each on its own stand, except the top two sections share a stand, as shown in Fig. 7.14. Played by one or more percussionists, the tubes are stuck with yarn-wrapped mallets or bowed with a bass bow. The tuning of the zoomoozophone follows the "tonality diamond" of Harry Partch (Partch instruments are described in Chapters 14 and 15) centered on G_4 (392 Hz).

Fig. 7.14. Zoomoozophone of Dean Drummond consists of 129 aluminum tubes tuned to a 31-note per octave scale (photo by Steve Hockstein).

7.10. African Lamellaphones: Mbira, Kalimba, Likembe, Sanza, Setinkane

Plucked metallophones or lamellaphones occur all over Africa and are known by such names as mbira, kalimba, likembe, sanza, etc. These instruments consist of sets of cantilevered tuned tongues fitted to a box or gourd resonator or a sounding board. Rattling devices are frequently attached to the resonator or sounding board to the tongues themselves.

Actually, "plucked" is not totally accurate. The free ends of the tongues are played by depressing them with the thumbs or other fingers and releasing them; hence the name "thumb piano" is sometimes applied to them. The tuning arrangement of the keys differs among the various types of instruments and in different regions. Lamellaphones are often used to accompany songs, although solo instrumental and ensemble music is also heard as well. Examples of African lamellaphones are shown in Fig. 7.15.

Metallophones

Fig. 7.15. African lamellaphones: (a) Mbira; (b) Kalimba; (c) Likembe (front and rear views).

Ethnomusicologist Andrew Tracey, who has studied many types of African lamellaphones, has concluded that the 8-note kalimba is the ancestor of most other instruments. The kalimba, he concludes, is at least 1000 years old and perhaps much older. A version with 8 to 14 keys, played with the thumbs and held over a small gourd, is played to the north of the Zambezi River, while a larger type, with 14 to 25 keys and held inside a large calabash, is more common south of the Zambezi River.[10]

Mbira are common in Zimbabwe and the lower Zambezi region in southeast Africa. The Shona-speaking people in southern Africa are fond of a large mbira with 22 or 23 lamellae called *mbira dza vadzimu* shown in Fig. 7.11 (a). In one playing style, described as *kukwenya*, the thumbs play in the normal manner while the right index finger scratches the six reeds in its playing area in an upward movement [11].

Likembe is the most common name given to a lamellaphone with a box resonator of Congolese (Zairean) origin, as shown in Fig. 7.11(c); other terms are kembe, dikembe, ikembe, and llukeme. The box has two sound holes, one on the rear and one on the end, which the player alternately opens and closes to produce vibrato and "wow" effects. Metal rings are generally added to the lower-pitched keys to produce a buzzing sound.

The *setinkane* is a lamellaphone of the Tswana people in Botswana. The tongues, once made of flattened nails, now are generally cut from sheet metal. The wooden soundboard is typically about 20 cm (8 in) long and 10 cm (4 in) wide. A buzzing sound was provided by placing a soft tissue from the bubu tree across a hole in the soundboard, although today it is more common to attach small metal beads to the sound board to create the buzzing sound. A setinkane is shown in Fig. 7.16.

Fig. 7.16. Setinkane played by a Bushman in Botswana (photo by Sarah S. Walters).

References

1. N. H. Fletcher and T. D. Rossing, *The Physics of Musical Instruments,* 2nd ed. (Springer-Verlag, New York, 1991).
2. T. D. Rossing, *The Physics Teacher* 14 (1976) 546.
3. R. S. Brindle, *Contemporary Percussion* (Oxford University Press, London, 1970).
4. K. Peinkofer and F. Tannigel, *Handbook of Percussion Instruments,"* English translation by K. and E. Stone (Schott, London, 1976).
5. M. Henderson, *Acoustics Australia* 22(1) (1984) 12.
6. J. I. Dunlop, *Acustica* 55 (1984) 250>
7. W. P. Malm, *Music Cultures of the Pacific, the Near East, and Asia* (Prentice-Hall, Englewood Cliffs, NJ, 1967) 25.
8. T. D. Rossing and R. B. Shepherd, *Percussive Notes* 19(3) (1982) 73.
9. L. Ayers and A. Horner, *J. Audio Eng. Soc.* **47** (1999), 813.
10. A. Tracey, *African Music* 5(2) (1972) 85.
11. R.A.Kauffman, G. Kubik, A. King, and P. Cooke, *New Grove Dictionary of Musical Instruments,* (Macmillan, London, 1984) 497.

Chapter 8
Interlude: Vibrations of Plates and Shells

In all musical instruments, the sound is generated and radiated by one or more simple or complex vibrating systems. Thus we considered the vibrations of strings and membranes in Chapter 2, the vibrations of bars, rods, and air columns in Chapter 5, and the vibrations of thick bars and tubes in Chapter 7. In this chapter, the vibrations of plates and shells will be considered.

8.1. Waves in a Thin Plate

A plate may be likened to a two-dimensional bar or beam in that it vibrates without the externally applied tension required to supply the restoring force in a string or membrane. Like a bar, it can transmit compressional waves, shear waves, torsional waves, or bending waves; and it can have three different boundary conditions: free, clamped, or simply-supported (hinged).

A plate might be expected to transmit longitudinal (compressional) waves at the same speed as a bar: $c_L = \sqrt{E/\rho}$. This is almost, but not quite, the case, however, since the slight lateral expansion that accompanies a longitudinal compression is constrained in the plane of the plate, thus adding a little additional stiffness. The correct expression for the velocity of longitudinal waves in an infinite plate is

$$c_L = \sqrt{\frac{E}{\rho(1-v^2)}},$$

where v is Poisson's ratio ($v \approx 0.3$ for most materials).[1]

Transverse waves in a solid involve mainly shear deformations. In a circular rod, transverse waves propagate at the same speed as torsional waves ($c_T = \sqrt{G/\rho}$). The shear modulus G is considerably smaller than Young's modulus E, however, so transverse and torsional waves propagate at roughly 60% of the speed of longitudinal waves ($c_L = \sqrt{E/\rho}$).

Bending waves in a plate are rather highly dispersive; that is, their speed $c(f)$ depends on frequency:

$$c(f) = \sqrt{1.8fhc_L}$$

where h is the thickness of the plate and c_L is the speed of longitudinal waves. The frequencies of the normal modes, of course, depend on the boundary conditions of the plate (shape and whether the edges are free, clamped, or simply-supported).

8.2. Circular Plates

The mathematical solutions of the equation of motion for vibrations of a circular plate, which are written as combinations of ordinary and hyperbolic Bessel functions, are beyond the scope of this book. The vibrational modes of circular plates are shown in Fig. 8.1. Note that the (0,0) and (1,0) modes are missing in a free plate; since the edge is not constrained there would be no restoring force, and displacements would lead to pure translational and rotational motion, respectively.

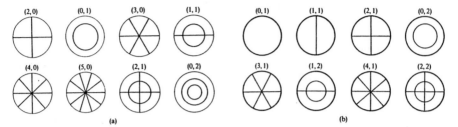

Fig. 8.1. Vibrational modes of circular plates: (a) free edge; (b) clamped or simply-supported edge. Frequency ratios are given in Tables 8.1 and 8.3.

In Tables 8.1-8.3 are the vibration frequencies of a circular plate with a free, clamped, and simply-supported edge. The mode number (n,m) gives the number of nodal diameters and nodal circles. The fundamental mode frequency is given in terms of the longitudianl wave speed c_L, thickness h, and radius a, and the remaining modes are given as ratios to the fundamental.

Table 8.1. Vibration frequencies of a circular plate with a free edge

—	—	$f_{20} = 0.2413 c_L h/a^2$	$f_{30} = 2.328 f_{20}$	$f_{40} = 4.11 f_{20}$	$f_{50} = 6.30 f_{20}$
$f_{01} = 1.73 f_{20}$	$f_{11} = 3.91 f_{20}$	$f_{21} = 6.71 f_{20}$	$f_{31} = 10.07 f_{20}$	$f_{41} = 13.92 f_{20}$	$f_{51} = 18.24 f_{20}$
$f_{02} = 7.34 f_{20}$	$f_{12} = 11.40 f_{20}$	$f_{22} = 15.97 f_{20}$	$f_{32} = 21.19 f_{20}$	$f_{42} = 27.18 f_{20}$	$f_{52} = 33.31 f_{20}$

Table 8.2. Vibration frequencies of a circular plate with a clamped edge

$f_{01} = 0.4694 c_L h/a^2$	$f_{11} = 2.08 f_{01}$	$f_{21} = 3.41 f_{01}$	$f_{31} = 5.00 f_{01}$	$f_{41} = 6.82 f_{01}$
$f_{02} = 3.89 f_{01}$	$f_{12} = 5.95 f_{01}$	$f_{22} = 8.28 f_{01}$	$f_{32} = 10.87 f_{01}$	$f_{42} = 13.71 f_{01}$
$f_{03} = 8.72 f_{01}$	$f_{13} = 11.75 f_{01}$	$f_{23} = 15.06 f_{01}$	$f_{33} = 18.63 f_{01}$	$f_{43} = 22.47 f_{01}$

Table 8.3. Vibration frequencies of a circular plate with a simply-supported (hinged) edge

$f_{01} = 0.2287 c_L h/a^2$	$f_{11} = 2.80 f_{01}$	$f_{21} = 5.15 f_{01}$
$f_{02} = 5.98 f_{01}$	$f_{12} = 9.75 f_{01}$	$f_{22} = 14.09 f_{01}$
$f_{03} = 14.91 f_{01}$	$f_{13} = 20.66 f_{01}$	$f_{23} = 26.99 f_{01}$

It is interesting to note that the fundamental mode in a circular plate with a simply-supported (hinged) edge has a lower frequency than the same plate with a free edge. This may be easily understood by noting that the center of mass moves up and down in the simply-supported plate but not in the free plate. Comparing modes of similar shape [the (2,0) and (0,1) modes in Fig. 8.1(a) with the (2,1) and (0,2) modes in Fig. 8.2(b), for example], however, the frequencies of the free plate modes are considerably smaller.

8.3. Elliptical Plates

The frequencies of an elliptical plate of moderately small eccentricity with a clamped edge are given approximately by the formula [1]

$$f \approx \frac{0.291 c_L h}{a^2} \sqrt{1 + \frac{2}{3}\left(\frac{a}{b}\right)^2 + \left(\frac{a}{b}\right)^4},$$

where a and b are the semimajor and semiminor axes. An elliptical plate with $a/b = 2$ has frequencies 37% greater than a circular plate with the same area.

8.4. Rectangular Plates

Since each edge of a rectangular plate can be free, clamped, or simply-supported, there are 27 different boundary conditions, and each leads to a different set of vibrational modes. Generally, the ones with similar conditions at the four edges are most interesting.

The mode frequencies for a plate with simply-supported edges, having dimensions L_x and L_y and thickness h are:

$$f_{mn} = 0.453 c_L h \left[\left(\frac{m+1}{L_x}\right)^2 + \left(\frac{n+1}{L_y}\right)^2\right].$$

where m and n are the numbers of nodes in the y and x directions, and c_L is the speed of longitudinal waves in the material. The nodal lines are parallel to the edges, whereas in plates with free or clamped edges, they are not.

Calculating the modes of a rectangular plate with free edges was described by the great Lord Rayleigh as a problem "of great difficulty." However, Rayleigh's own methods lead to approximate solutions that are close to measured values, and refinements by Ritz bring them even closer. Results of many investigations are summarized by Leissa [2].

Actually, the vibrational modes of a free plate would be similar to those of a thin bar with free ends, given in Section 5.1, but for one thing. As the bar takes on appreciable width, bending along one axis causes bending in a perpendicular direction. This comes about because the upper part of the bar above the neutral axis becomes longer (and thus narrower), while the lower part becomes shorter (and thus wider). In fact, Poisson's constant v is a measure of the lateral contraction that accompanies a longitudinal expansion of a plate, and the factor $1-v^2$ appears in the expressions for both longitudinal and bending wave speeds.

Several bending modes in a rectangular plate can be derived from the bending modes of a bar. The $(m,0)$ modes might be expected to have nodal lines parallel to one pair of sides, and the $(0,n)$ would have nodal lines parallel to the other pair of sides. Because of the coupling between bending motions in the two directions, however, the modes are not pure bar modes. The nodal lines become curved, and the plate takes a sort of saddle shape (i.e., concave in one direction but convex in the perpendicular direction). This can be called anticlastic bending, and it is quite evident in the modes of two different rectangular plates shown in Fig. 8.2. Chladni patterns, made by sprinkling sand, salt, or powder on a vibrating plate, are useful for viewing the modes of vibration of plates as well as membranes (see Section 2.9). The powder collects at the nodes.

Modes in the rectangular plates in Fig. 8.2 can be designated by the numbers of vertical and horizontal nodal lines; the (m,n) mode has m vertical nodal lines and n horizontal nodal lines. The m-numbers appear at the top of the figure and the n-numbers at the left.

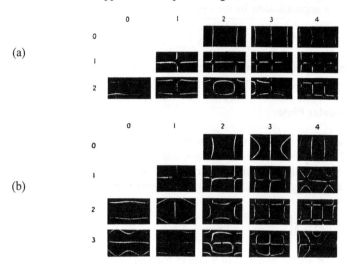

Fig. 8.2. Chladni patterns showing the vibrational modes of rectangular plates of different shapes: (a) $L_x/L_y=2$; (b) $L_x/L_y=3/2$ [4].

Relative frequencies for the plate with $L_x/L_y=2$ in Fig. 8.2 are given in Table 8.4.

Table 8.4. Relative frequencies of modes in a free-edge plate with $L_x/L_y=2$ [4]

	$m=0$	1	2	3	4
$n=0$			1	2.88	5.42
1		1.20	2.30	3.62	6.2
2	4.37	4.87	6.7	8.2	10.8

It is interesting to note how the combinations develop in a rectangle as L_{xc}/L_y approaches unity. Fig. 8.3 shows the shapes of two modes that are descendants of the (2,0) and (0,2) beam modes in a rectangle of varying L_x/L_y. When $L_x \gg L_y$, the (2,0) and (0,2) modes appear quite independent. However, as $L_x \to L_y$, the beam modes mix together to form two new modes. In the square, the mixing is complete and two combinations are possible, depending upon whether the component modes are in phase or out of phase. For a further discussion of the modes in square plates and in plates of anisotropic material such as wood, see Chapter 3 in reference [1].

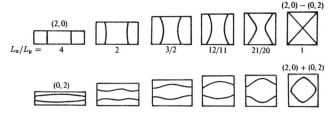

Fig. 8.3. Mixing of the (2,0) and (0,2) modes in rectangular plates with different L_x/L_y ratios (after Waller [5]).

8.5. Cylindrical Shells

A shell is generally defined as a curved surface having dimensions large compared with its thickness. Rayleigh identified two types of vibrational modes, which he called extensional and inextensional (or flexural). In an *extensional* mode, a line drawn on the shell surface changes its length during vibration, and the elastic forces associated with this change are quite large. To a first approximation, the potential energy due to the deformation and the kinetic energy are both proportional to the thickness h and the modal frequencies are independent of h. In a *flexural* mode, on the other hand, the potential energy is proportional to h^3 and the frequency is proportional to h. For modes that involve both extensional and inextensional deformations, the frequency can be written as $f_{mn}=(A_{mn}+B_{mn}h^2)^{1/2}$, where A_{mn} and B_{mn} are appropriate constants that include the dimensions, the type of material, etc.[1].

For a thin circular ring, the lowest extensional mode is a simple "breathing" mode, as shown in Fig. 8.4(a). The frequency of this mode is $f_{oE} = 1/(2\pi a)\sqrt{E/\rho}$, where a is the radius, E is the elastic (Young's) modulus, and ρ is the density. The frequency is inversely proportional to the radius but does not depend on the thickness of the ring (just as the longitudinal vibration frequencies of a bar or rod are inversely proportional to its length but independent of its thickness; see Section 5.2).

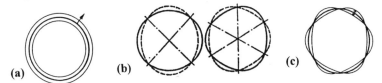

Fig. 8.4. Vibrations of a ring: (a) Lowest extensional mode; (b,c) Lowest two inextensional (flexural) modes.

The lowest two inextensional (flexural) modes of a ring are shown in Fig. 8.4(b,c). Note that there is both radial and tangential motion (there has to be if the mode is inextensional). The radial velocity is given by $u=A_m m \sin m\theta$ and the tangential velocity by $v=A_m \cos m\theta$. Thus, there are no true nodes, although we sometimes speak of the $2m$ points that have no radial motion as "nodes." Note that the tangential velocity is m times smaller than the radial velocity, so it becomes increasingly insignificant as m increases.

A circular cylinder can be thought of being made up of ring-shaped elements, each of which can move in the manner just described. The $2m$ nodes of the ring become nodal lines parallel to the axis of the cylinder, and in addition, there are n nodal circles. In addition to the inextensional modes shown in Fig. 8.4, there are also inextensional modes with $m=0$ and $m=1$, as shown in Fig. 8.5(b) (in a ring these would be interpreted as rotational and translational rather than vibrational motion). Higher extensional modes are shown in Fig. 8.4(a).

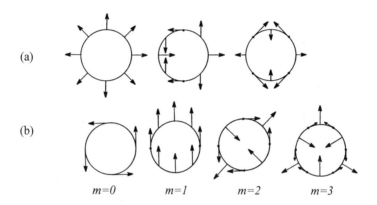

Fig. 8.5. Modes of a circular cylinder: (a) extensional; (b) inextensional.

In a circular cylinder with free ends, the extensional and inextensional mode frequencies for $n=0$ are the same as in a circular ring:

$$f_{mE} = 1/2\pi a\sqrt{E/\rho}\,(m^2+1)$$

$$f_{ml} = h/2\pi a^2 \sqrt{E/12\rho}\, m(m^2-1)/\sqrt{m^2+1}$$

where a is the radius, h is the thickness, E is the elastic (Young's) modulus, and ρ is the density. The factor $m(m^2-1)/\sqrt{m^2+1}$ can be approximated by m^2-1 and eventually by m^2 as m gets larger. Note that $f_{ml}=0$ for $m=0$ and $m=1$, since these represent pure rotational and translation motion as long as $n=0$. We now consider the more general case where $n>0$.

Modes of vibration in which $m>0$ and $n>0$, so that there are both circumferential and axial nodes, are no longer inextensional, and the analysis becomes more complicated. Both bending and stretching of the shell must be considered. The (1,1) mode, for which $m=n=1$ is essentially a bending vibration of the shell, as shown in Fig. 8.6(a). There are two nodal circles and two axial nodal lines. This is the fundamental mode of vibration in an orchestra bell or chime (Section 7.5). The (2,0) mode ($m=2$, $n=0$) is shown in Fig. 8.6(b) for comparison. In general, the (m,n) mode will have both nodal circles and axial nodal lines, as shown in Fig. 8.6(c).

Vibrations of Plates and Shells 85

Fig. 8.6. Modes of a circular cylinder: (a) (1,1) mode; (b) (2,0) mode; (c) (m,n) mode with n nodal circles and $2m$ axial nodal lines.

Calculating the modal frequencies for modes with $n>0$ becomes rather complicated. Fig. 8.7 shows one example, for the case in which $a/h=78$, $L/a=2.4$. For $n=0$ (curve labeled "inextensional") the frequency increases with increasing m, but for all the other curves, the frequency passes through a minimum frequency at some value of m, which is somewhat surprising at first glance. This is due the fact that the total strain (potential) energy is made up of both bending and stretching energy. For large m-values, the bending energy is large, but for small m-values the stretching energy tends to increase with decreasing m.

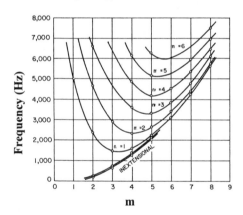

Fig. 8.7. Vibrational mode frequencies in a circular shell with free ends with n nodal circles and $2m$ axial nodes [6].

The ends of cylindrical shells can be constrained in quite a number of ways. They can be rigidly clamped, they can be simply supported, they can be terminated by a "shear diaphragm (which forces the cross section to remain circular but permits other motion), for example. Books and papers on elasticity theory consider as many as 136 different combinations of end conditions. Fig. 8.8(a) shows experimental and theoretical frequencies for a circular cylinder with simply-supported ends, and Fig.8.8(b) shows the contributions to strain energy from bending and stretching in this cylinder [7].

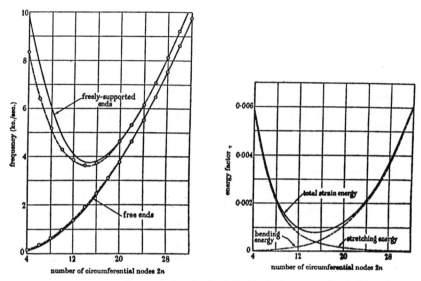

Fig. 8.8. (a) Vibrational mode frequencies in a circular shell with simply-supported ends; (b) strain energy due to bending and stretching [7].

Fig. 8.9 shows the vibrational frequencies in a circular shell with simply-supported ends having a/h=38.1 and L/a=7.9. When logarithmic axes are used for plotting f and m, the n=0 data nearly follow a straight line, while the n>1 curves show a minimum.

Fig. 8.9. Vibrational mode frequencies in a circular shell with simply-supported ends having a/h=38.1 and L/a=7.9.

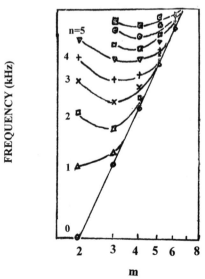

8.6. Shallow Spherical Shells

Arching a plate adds greatly to its stiffness, just as an arch in a building structure adds greatly to its rigidity. Arching the top plate of a violin adds enough static stiffness to withstand the considerable downward force of the strings on the bridge, for example, and it also adds vibrational stiffness. The vibrational frequency of a circular plate is doubled if the arch height is a little more than twice the thickness. For a shell of thickness h with a rise H at the center, Reissner [8] shows that the frequency ratio of the lowest axisymmetric mode of a shell (which is actually an extensional mode) to that of a comparable flat plate rises to $0.68H/h$ for a shell with a clamped edge having $H/h>>1$ or $0.84H/h$ for a shell with a free edge.

When the shell is thin enough that H/h is greater than about 20 and shallow enough that $H/a<0.25$, the lowest mode frequency is approximately $f \approx H/(\pi a^2)\sqrt{E/\rho}$ whether the shell is clamped or free at its edge. The higher modes are less affected by shell curvature than the fundamental mode [9].

8.7. Nonlinear Effects in Plates and Shells

If we pull down on a mass supported by a spring and release it, the mass will vibrate in simple harmonic motion at a single frequency. The restoring force is proportional to the displacement until the spring is stretched beyond its elastic limit. A graph of force vs displacement shows the linear relationship, since a straight line results up to the elastic limit, at which time the line begins to curve and we observe a nonlinear relationship between force and displacement.

In strings and membranes, we observe nonlinearity at large amplitudes because the average tension on the string or membrane increases with amplitude. The increased tension causes a frequency increase which can be heard as a "twang" when a string is plucked with large amplitude or a pitch glide when a drum is struck a strong blow. Strings and membranes exhibit a "hardening spring" type of nonlinearity, since the vibrating system behaves as if the spring constant had increased at large amplitude.

Plates likewise show a hardening spring behavior but shallow spherical shells can show either a hardening spring behavior or a "softening spring" behavior, depending upon the geometry of the shell. Nonlinear effects in plates and shells have several applications in percussion musical instruments (such as the cymbals and gongs discussed in Chapter 9).

In a flat circular plate with thickness h, the frequency of the lowest mode of vibration varies with amplitude A approximately as $f=f_o[1+0.16(A/h)^2]$ where f_o is its frequency of vibration at small amplitude. The reason for the dependence on plate thickness is that this determines the relative importance of stiffness and tension forces in plate behavior [10].

A spherical cap shell of height H and thickness h has a rather complicated nonlinear behavior which is illustrated in Fig. 8.10. The frequency of the lowest (axisymmetric) mode depnds not only on the ratio A/H between the vibration amplitude and shell height but also on the geometrical quantity h/H. If the shell is very thin, so that $h/H<<1$, then nonlinear effects dominate the behavior. As the vibration amplitude is increased to approach H, the mode frequency falls to about half its small-amplitude value, while further increase in amplitude causes it to rise again. For progressively thicker shells, the nonlinearity is less important because of the dominance of bending stiffness, and very thick shells with $h>>H$ behave essentially as flat plates [10,11].

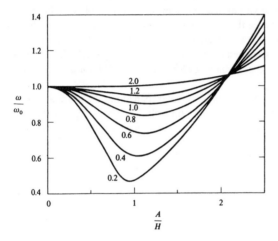

Fig. 8.10. Calculated ratio of frequency to small amplitude frequency for lowest axisymmetric mode of a spherical-cap shell of dome height H and thickness h when vibrating with amplitude A. Values of h/H are shown for each curve [11].

References

1. N. H. Fletcher and T. D. Rossing, *The Physics of Musical Instruments, 2nd ed.* (Springer-Verlag, New York, 1991). Chapter 3.
2. A. W. Leissa, *Vibration of Plates,* (NASA, Washinton, D.C., 1969) (Reprinted by Acoustical Society of America, Woodbury, NY, 1993).
3. E. F. F. Chladni, *Die Akustik,* 2nd ed. (Breitkofp u. Härtel, Leipzig, 1802).
4. M. D. Waller, *Proc. Phys. Soc. London* **B62** (1949) 451.
5. M. D. Waller, *Chladni Figures: A Study in Symmetry.* (Bell, London, 1961).
6. A. W. Leissa, *Vibration of Shells* (NASA, Washington, D.C., 1973). (Reprinted by Acoustical Society of America, Woodbury, NY, 1993).
7. R. N. Arnold and G. B. Warburton, *Proc. Roy. Soc. (London)* **A197** (1953) 238.
8. E. Reissner, *Q. Appl. Math.* **13** (1955) 279.
9. A. Kalnins, *Proc. 4th U.S. Cong. Appl. Mech.* (1963) 225.
10. N. H. Fletcher and T. D. Rossing, *The Physics of Musical Instruments, 2nd ed.* (Springer-Verlag, New York, 1991). Chapter 5.
11. N. H. Fletcher, *J. Acoust. Soc. Am.* **78** (1985) 2069.

Chapter 9
Cymbals, Gongs, and Plates

This chapter considers idiophones that are based on vibrating plates and shallow spherical shells, two types of vibrators discussed in Chapter 8. These idiophones are generally set into vibration by striking with a mallet or beater, but sometimes sustained tones are produced by bowing the edges with bass or cello bows.

9.1. Cymbals

Cymbals are among the oldest of musical instruments and have had both religious and military use in a number of cultures. Cymbals range from 20 cm to 75 cm (8 to 30 in) in diameter and are generally made of bronze. The Turkish cymbals generally used in orchestras and bands are saucer-shaped with a small dome in the center, in contrast to Chinese cymbals, which have a turned-up edge more like a tamtam. Given the wide ranges in diameter and thickness, considerable variety of cymbal tone is available to the percussionist.

Many different types of cymbals are used in orchestras, marching bands, concert bands, and jazz bands. Orchestral cymbals are usually between 40 and 55 cm (16 to 22 in) in diameter and are often designated as "French," "Viennese," and "Germanic" in order of increasing thickness. Jazz drummers use cymbals designated by such onomatopoeic terms as "crash," "ride," "swish," "splash," "ping," and "pang."

A good cymbal can be made to produce many different tones by using a variety of sticks and striking it at several different places. A large cymbal struck gently near the rim produces a low sound not unlike that of a small tamtam. The fullest sound is obtained by a glancing blow about one-third of the way in from the rim.

At one time, orchestral cymbals were always used in pairs, clashed or crashed together (hence the term "crash" cymbals), although now they are played in other ways as well. Orchestral players take great care in choosing crash cymbals, generally preferring that one have a slightly lower "pitch" than the other [1].

9.2. Vibrational Modes of Cymbals

Using electronic TV holography, Wilbur [2] recorded over 100 modes of vibration in a 46-cm diameter medium crash cymbal, 23 of which are shown in Fig. 9.1. The bright lines indicate the nodal lines and the fringes indicate isoamplitude contours. The edge, which lies just outside the largest nodal circle, is difficult to see. The modes are labeled (m,n), the first number m designating the number of nodal diameters and the second number n the number of nodal circles. In the $n=0$ modes (top row), the center of the cymbal vibrates very little, and this nodal region expands as m increases. The same 23 modes in a flat plate of the same diameter are shown in Fig. 9.2 for comparison. In the flat plate, the vibrations in the $n=0$ modes extend closer to the center, especially in the $(2,0)$ and $(3,0)$ modes. Both the cymbal and the plate were supported at the center in these studies [2].

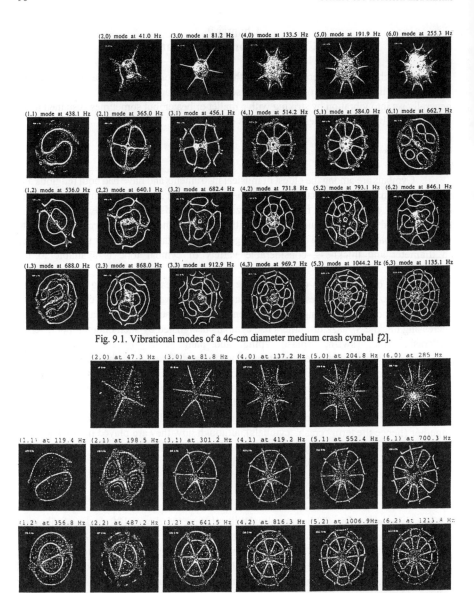

Fig. 9.1. Vibrational modes of a 46-cm diameter medium crash cymbal [2].

Fig. 9.2. Vibrational modes of a 46-cm diameter flat plate [2].

In a circular plate, the modal frequencies of the $n=0$ modes (those without nodal diameters) appear to follow a relationship $f_{m,o}=Cm^2$, in other words, they are proportional to the square of the number of nodal diameters. Chladni [3] observed that the addition of one nodal circle raised the frequency of the plate by about the same amount as adding two nodal diameters, leading to a relationship $f_{m,n}=C(m+2n)^2$ that Lord Rayleigh [4] called Chladni's law.

We have found that modes in a flat circular plate are better described by modifying Chladni's law in one of two ways: $f_{mn}=C(m+3n)^p$ or $f_{mn}=C_n(m+2n)^p$ [5]. In the second case, different parameters C_n and p_n were used for each value of n. In the case of nonflat plates, such as bells and cymbals, it is found that further modification to the form $f_{mn}=C_n(m+bn)^p$ where b may vary from 2 to 4, is sometimes desirable.

Fig. 9.3 shows the modal frequencies of the 46-cm medium crash cymbal in Fig. 9.1 plotted as a function of number of modal diameters m. Each curve represents a different value of the number of nodal circles n. We have set $b=3$ (i.e., have used the form $m+3n$) and values of p_n range from 1.56 to 1.28 [2].

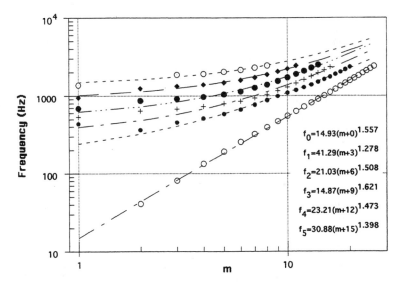

Fig. 9.3. Modal frequencies of a 46-cm-diameter medium crash cymbal as a function of m and n, illustrating Chladni's law [2].

In Fig. 9.3 it appears that to best fit the $n=0$ family of modes to Chladni's law, a different value of p should be used at small and large m. In most cymbals, a distinct change in slope is noted at some particular value of m, which is denoted as m_c in Table 9.1. The largest p-values occur when the cymbal is large and thin; the cymbal then approaches flat-plate behavior ($p=2$).

Table 9.1. Parameters used to fit the vibration modes of cymbals having $n=0$ to the equation $f=c(m+2n)^p$ [5]

Cymbal	p_1	c_1 (Hz)	p_2	c_2 (Hz)	m_c
24 in. thin	1.86	7.7	1.49	14.2	6
18 in. thin	1.75	10.1	1.56	14.1	5
	1.78	10.6	1.52	15.7	4.5
18 in. medium	1.65	13.4	1.46	18.2	4.7
	1.70	12.6	1.43	17.8	3.6
16 in. thin	1.81	10.8	1.48	18.8	5
	1.84	12.0	1.47	19.5	4
16 in. medium	1.70	13.8	1.53	18.3	5
	1.65	15.9	1.53	19.4	4.3
15 in. thick	1.47	20.6			

9.3. Cymbal Sound

At least three prominent features have been observed in the sound of a cymbal: the strike sound that results from rapid wave propagation during the first millisecond, the buildup of strong peaks around 700-1000 Hz in the sound spectrum during the next 10 or 20 ms, and the strong aftersound in the range of 3-5 kHz that dominates the sound a second or so after striking and gives the cymbal its "shimmer."

The propagation of bending waves on a cymbal immediately after momentary excitement with a laser pulse can be traced, as shown in Fig. 9.4, by using pulsed video holography [6]. The field of view is about 20 x 15 cm and the cymbal was pulsed at a point 10 cm (about half its radius) from the outer edge. Because of dispersion of bending waves (see Section 8.1), the first to be seen have short wavelengths, about 5 mm in this case, a frequency of about 340 kHz and a propagation speed of about 1700 m/s. These are followed by waves of longer wavelength and greater amplitude, which are subsequently reflected from the outer edge of the cymbal (at the far right-hand side) and from the central dome (near the left-hand side) to interfere with the circular wave fronts. The normal modes do not, however, become distinguishable until a time Δt after the initial excitation that is determined by the separation Δf between the modes by the relation $\Delta f \Delta t \approx 1$.

Several years ago we measured the decay times of single modes of vibration and also octave bands of sound when the cymbal is struck. In order to measure the decay times of single modes, the mode of interest is excited with a magnetic driver, which is then switched and the sound level recorded as the vibration decays. Decay times (60 dB) for nine radial modes in a 40-cm medium crash cymbal are shown in Fig. 9.5. Also shown in Fig. 9.5 are decay times of the octave bands of sound when the cymbal is struck in three different ways: a crash stroke (C) in which the shoulder of a wooden drumstick contacts the cymbal near the edge with a hard, glancing blow; a ride stroke (R) in which the plastic tip of the stick impacts the cymbal about halfway between the edge and the dome of the cymbal; and a soft stroke (S) near the edge with a timpani mallet. The sound from the ride stroke appears to be more sustained in the lower octave bands than from the crash stroke, but the reverse is true at 16 kHz [7].

Cymbals, Gongs, and Plates

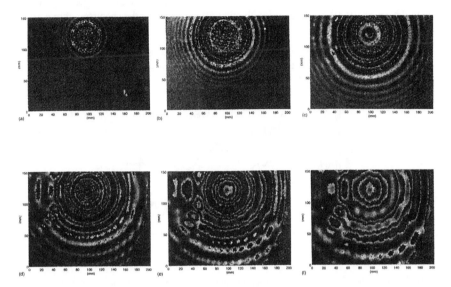

Fig. 9.4. Phase maps showing wave propagation outward from a point 10 cm from the edge of a cymbal. Time after impulse: (a) 30 μs; (b) 60 μs; (c) 120 μs; (d) 180 μs; (e) 240 μs; (f) 300 μs [6].

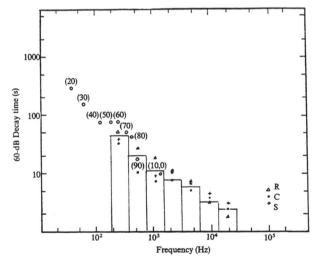

Fig. 9.5. Decay times of single modes of vibration and octave bands of sound in a 400-cm-diameter medium crash cymbal. Octave-band decays were measured for three different strokes (see text), designated as crash (C), ride (R), and soft (S) [7].

Sound spectra from the same cymbal at the time of striking and after intervals of 0.05 s, 1.0 s, and 2.0 s are shown in Fig. 9.6. Note that the sound level in the range of 2-10 kHz builds up by 10 dB or more during the first 50 ms, whereas the spectrum above 10 kHz shows little change.

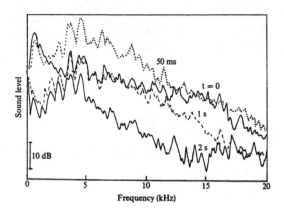

Fig. 9.6. Sound spectra of a 40-cm medium crash cymbal immediately after striking and after 0.05 s, 1.0 s, and 2.0 s [7].

From Fig. 9.6 and many other similar spectra, we can make the following observations:
1. The sound level below about 700 Hz shows a rather rapid decrease during the first 200 ms, after which it decays slowly. This is apparently caused by the conversion of energy into modes of higher frequency.
2. Several strong peaks in the 700-1000 Hz range build up between 100-20 ms, then decay.
3. Sound energy in the important 3-5 kHz range peaks about 50-100 ms after striking.
4. Sound in the range of 3-5 kHz, which gives the cymbal its "shimmer," is often the most prominent feature from about 1-4 s after striking.
5. The low frequencies again dominate the lingering sound, but at a much lower level, so that they are rather inconspicuous.

9.4. Nonlinear Behavior of Cymbals

The conversion of energy from the low-frequency modes that are initially excited when the cymbal is struck into the high-frequency vibrations that are responsible for much of the characteristic cymbal sound embodies some interesting physics. There is considerable evidence that the vibrations exhibit chaotic behavior [8,9,10,11]. The road to chaos appears to follow the following stages: first the generation of harmonics, then the generation of subharmonics, and finally chaotic behavior.

Cymbals, Gongs, and Plates

If a flat circular plate is excited sinusoidally, it vibrates in the normal modes of Fig. 9.3 at low amplitudes. Large-amplitude excitation at a frequency near one of these modes leads to "bifurcation" associated with doubling or tripling of the vibration period. The same behavior is observed for an orchestral cymbal excited sinusoidally at its center, one particular cymbal giving a five-fold increase in period and a subsequent major-chord-like sound based on the fifth subharmonic of the excitation frequency [9]. The fourth and seventh subharmonics have been observed in other cymbals [10], along with harmonics of these subharmonics (in other words, partials having frequencies $n/4$, $n/5$, or $n/7$ times the excitation frequency).

Figure 9.7 shows phase plots (velocity vs displacement near the edge) for a Zildjian thin crash cymbal 41 cm (16 inches) in diameter driven sinusoidally at the center with a shaker. The driving frequency was 192 Hz and the drive amplitude started small and increased in 3 steps over a 20:1 range. Note the successive appearance of harmonics, subharmonics, and chaotic behavior.

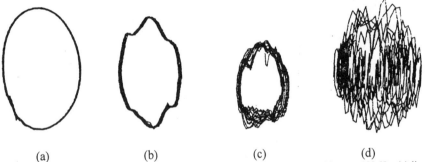

(a) (b) (c) (d)

Fig. 9.7. Phase plots (velocity vs displacement) for a thin crash cymbal center driven at 192 Hz: (a) linear behavior at 0.05 A; (b) harmonics present at 0.15 A; (c) subharmonics (including their harmonics) present at 0.5 A; (d) chaotic vibration at 1 A [10].

Velocity spectra and phase plots are shown in Fig. 9.8 for the same cymbal center driven at 320 Hz. The first set shows harmonics of the driving frequency, while the second set at twice the driving current shows subharmonic generation, and the third shows chaotic behavior. In the spectrum of Fig. 9.8(b), we can see harmonics of 1/5 the driving frequency, four of which appear between successive harmonics of the driving frequency (marked by small squares). When the drive frequency was raised to 450 Hz, the road to chaos was similar, but the subharmonics are now harmonics of 1/7 the driving frequency [10].

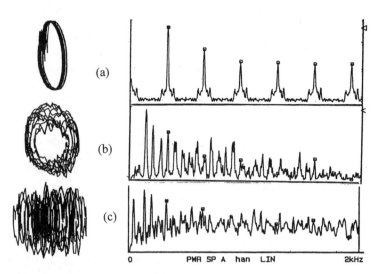

Fig. 9.8. Phase plots and velocity spectra for a cymbal center drive at 320 Hz: (a) harmonic generation at 0.3 A drive current; (b) subharmonics at 0.6 A; (c) chaotic behavior at 1.4 A [10].

Brass, bronze, and steel plates of the same size (and approximately the same thickness) as the cymbal also showed nonlinear behavior leading to chaos. In general, subharmonic generation and chaotic behavior required slightly higher vibration amplitudes in the flat plates than in the cymbals. Subharmonic generation was more difficult to observe in the plates; in general, they tended to move directly to chaotic behavior as the vibration amplitude increased.

A mathematical analysis of cymbal vibrations using nonlinear signal processing methods reveals that there are between 3 and 7 active degrees of freedom and that physical modeling will require a like number of equations [11]. Because of its nonlinear behavior, the cymbal is a difficult, but not impossible, musical instrument to model for sound synthesis, for example. In fact, percussionists are sometimes asked to play live "cymbal sounds" for recordings using electronic drum machines.

9.5. Tam-Tams

Chinese tam-tams, widely used in symphony orchestras, are not only among the loudest instruments in the orchestra, they are also among the most scientifically interesting. When the tam-tam is struck somewhere near its center with a large padded mallet, the initial sound is one of very low pitch, but in a few seconds a louder sound of high pitch builds up, then slowly decays, leaving once again a lingering sound of low pitch. The high-pitched sound fails to develop if the initial blow is not hard enough.

Tam-tams are of varying size, generally up to about 1 m in diameter. They are usually made of bronze (approximately 80% copper and 20% tin, with traces of lead or iron). Although the center is usually raised slightly, they do not have a prominent central dome as do gongs and cymbals. They do, however, have one or more circles of hammered bumps and a fairly deep rim. They are thinner than most large gongs. Tam-tams are widely used for creating special effects, such as in Moussorgsky's "Night on Bald Mountain" or the sound produced by the Herculean figure in the familiar screen trademark of the J. Arthur Rank Film Corporation [12].

The modes of vibration of a large tam-tam has several axisymmetric modes of low frequency that tend to absorb much of the energy of the initial blow. In one large tam-tam of Japanese manufacture, for example, modes with frequencies of 39, 162, 195, 318, 854, and 1000 Hz were observed when the tam-tam was driven at its center [13]

The buildup and decay of vibrations at various frequencies during the first 0.4 s after vigorous excitation of a tam-tam is shown in Fig. 9.9. These waveforms were recorded with an accelerometer placed approximately halfway between the center and edge. The delay in excitation of the higher modes is clearly apparent, as well as the irregular amplitude variation in each frequency band. The most likely explanation for this is that the vibration is chaotic, as we have already discussed for the cymbal. Since the frequency band accepted by the analyzer in this measurement was only 100 Hz, such chaotic vibration would be expected to show rather wide amplitude fluctuations.

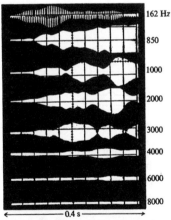

Fig. 9.9. Buildup and decay of vibrations in different frequency bands during the first 0.4 s [13].

Sound spectra immediately after excitation and again after 3 s are shown in Fig. 9.10. The large initial excitation of low-frequency modes is apparent, as is the subsequent transfer of vibrational energy to modes in the range 1-5 kHz which contributes to the late developing shimmer. The relatively slow buildup of high-frequency modes gives the tam-tam its distinctive timbre. The nature of the nonlinear coupling between the modes is not well understood at present, but the large number of hammered bumps spaced around the tam-tam appear to play a significant role in transferring energy from axisymmetric modes to modes of lower symmetry. The harder the blow, the greater the nonlinear coupling [14].

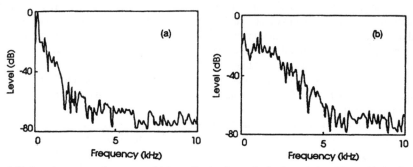

Fig. 9.10. Sound spectrum of a tam-tam: (a) immediately after excitation by a stroke with a padded striker; (b) after 3 s [14].

9.6. Gongs

Gongs of many different sizes and shapes, a few of which are shown in Fig. 9.11, are used in both Western and Oriental music. They are usually cast of bronze with a deep rim and a protruding dome. Gongs used in symphony orchestras generally range from 0.5 to 1 m (20 to 38 in) in diameter, and emit sound with a rather definite pitch.

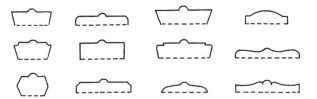

Fig. 9.11. Gongs of various shapes. Note the protruding dome at the center of each gong.

Nowhere is the gong revered more deeply than in Indonesia, where gongs of various sizes are the backbone of the gamelans of Java and Bali. A large gong is used to mark the end of each melodic section in gamelan music, for example. A large gong from Java and a small Balinese tawa tawa are shown in Fig. 9.12.

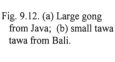

Fig. 9.12. (a) Large gong from Java; (b) small tawa tawa from Bali.

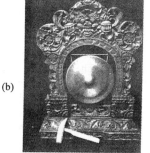

Sound spectra of the Balinese tawa tawa gong in Fig. 9.12(b) are shown in Fig. 9.13. Note that the initial sound comes almost entirely from two axially symmetric modes. Although they appear the same in the sketches, the one lower in frequency has a node where the face joins the rim, whereas the higher one has a node about one-third of the way down the rim. The second spectrum, recorded a half second after the gong is struck, shows that many other modes of vibration quickly develop and persist for a second or two. The top spectrum shows that 5 s after the strike the fundamental and a pair of modes with one nodal diameter dominate. Note that the two axisymmetric modes that produce the two prominent peaks in the spectrum at strike ($t=0$) have frequencies in a 2:1 (octave) ratio. We have found this to be the case in large gamelan gongs as well.

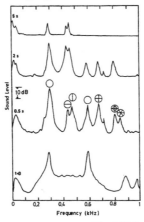

Fig. 9.13. Sound spectra of a Balinese tawa tawa gong. The initial sound ($t=0$) comes mainly from two prominent axisymmetric modes, but after 0.5 s many modes of vibration have been excited, which decay at varying rates. Some of the modes are identified at the peaks [15].

Figure 9.14 shows the sound spectrum of a large (59 cm in diameter) gamelan gong struck at its center. The frequencies of the two prominent axisymmetric modes are 67 Hz and 135 Hz, and the resulting pitch is identified as C_2. If the gong is struck off-center, a peak appears at 115 Hz, which is due to a mode with one nodal diameter. In this large gong, too, many modes develop a short time after the initial strike.

A common feature in both the large gamelan gong and the smaller tawa tawa gong is the octave interval between the two prominent partials in the initial sound. The mass of the central dome, compared to the rest of the gong, is apparently an important factor in determining the frequency ratio of the first two axisymmetric modes. Our best estimates indicate that the dome in the large gamelan gong has about 10% of the total mass of the gong, whereas in the smaller tawa tawa gong the dome mass is about 7% of the total.

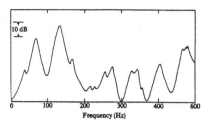

Fig. 9.14. Sound spectrum of a large gamelan gong. The principal modes of vibration have frequencies of 67 Hz and 135 Hz, so their corresponding partials are about an octave apart [15].

To study the tuning effect of the central dome mass, we performed a series of experiments on flat gongs that did not normally have an octave relationship in their spectra (and hence did not convey as strong a sense of pitch). We measured the frequencies of the modes of vibration as varying amounts of mass were bolted to the center of the gong.

The frequencies of the two lowest axisymmetric modes in a Zildjian Turkish gong are shown in Fig. 9.15. Although both modes decrease in frequency as the mass at the center increases, the lower one (which has an antinode at the center) falls faster than the higher one (which has its antinode about half way between the center and the rim). The octave interval in this gong occurs with a load of about 600 g at the center. With this load, the gong takes on an easily discernible pitch and a timbre not unlike that of a gamelan gong. The unloaded gong has a mass of 5300 g, so the critical load is about 10% of the total mass (5900 g), similar to that of the large gamelan gong.

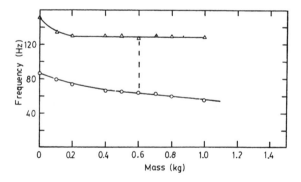

Fig. 9.15. The frequencies of the two principal axisymmetric modes of vibration in a Turkish gong loaded with various amounts of mass in the center. The dashed line at 600 g indicates the load that leads to a 2:1 (octave) ratio between the two modal frequencies.

9.7. Chinese Opera Gongs

Among the many gongs in Chinese music are a pair of gongs used in Chinese opera

Cymbals, Gongs, and Plates

orchestras, shown in Fig. 9.16. These gongs are of considerable scientific, as well as musical, interest because of their pronounced nonlinear behavior. The pitch of the larger gong glides downward as much as three semitones after striking, whereas that of the smaller gong glides upward by about two semitones [16].

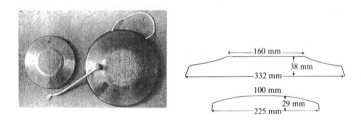

Fig. 9.16. Two Chinese opera gongs that show a marked pitch glide after being struck a substantial blow. The smaller gong at the left glides upward, while the larger gong at the right glides downward in pitch [16].

The shift in vibration frequency with amplitude is another example of the nonlinear effects in plates and shells discussed in Section 8.7. The central section of the larger gong is nearly flat, and the hardening spring behavior, characteristic of flat plates, dominates. The central part of the smaller gong, on the other hand, is sufficiently convex to behave as a spherical cap shell that has softening spring behavior at large amplitude. The sloping sides and the rim of the gongs normally vibrate relatively little, but they provide an edge condition that appears to be somewhere between clamped and simply-supported (hinged).

9.8. Bronze Drums

The bronze drum is one of the oldest continuous art traditions in Southeast Asia. The first bronze drums have been dated to the sixth century B.C. and were found in northern Vietnam. Bronze drums are important to the culture of the Karen people, who live mainly in Burma and the mountainous region between Burma and Thailand. Karen bronze drums date back at least to the sixteenth century, but they are believed to have had such drums long before this time. Karen people in Burma are a distinct minority ethnic group, and the separate Karen tribes do not necessarily speak the same languages or have the same customs.

A Karen bronze drum, which is cast in one piece except for the animal adornments, consists of an overlapping tympanum that may range from nine to thirty inches in diameter and a cylinder that is slightly longer than the diameter. Bronze drums have a wide variety of ritual use, both musical and non-musical.

The sound spectrum of a drum shows a dense collection of partials out to about 3 kHz. The corresponding modes of vibration, recorded using electronic TV holography, sometimes show the strongest vibration in the tympanum (as in the upper row in Fig. 9.17) and sometimes in the cylindrical shell (as in the bottom row). In all modes, the entire drum vibrates and radiates sound.

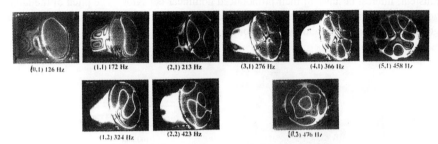

Fig. 9.17. Electronic TV holograms of vibrational modes in a Karen bronze drum [17].

The tympanum can be modeled as a circular plate, having either a clamped edge or a simply-supported (hinged) edge. The tympanum mode frequencies are in better agreement with those calculated for a flat brass plate with a simply-supported edge than for a plate with a clamped edge. The observed mode frequencies are somewhat less than the calculated frequencies, however, partly due to the mass of the overhanging portion.

Bronze drums were very important to the Karen people and to the Karen culture. They most likely did not construct their own drums, but had the Shan craftsmen of eastern Burma create the drums for them. The owner of a drum richly decorated with frogs was said to be more important to the community than if he owned seven elephants. Ownership of a frog drum was a sign of wealth and status in the Karen community, and village wars and raids occurred over the theft of one or more drums.

9.9. Crotales

Although the term crotales originally applied to small antique finger cymbals or castanets, modern composers use the term for different types of thick small cymbals, varying in shape from a small Turkish cymbal to a cup-like tuned bell played with a mallet. An octave set of the latter type is shown in Fig. 9.18. These are cupped bronze plates 4.5 mm thick and 101 to 129 mm in diameter.

Fig. 9.18. (a) A set of crotales tuned to notes on the diatonic scale from C_6 to C_7; (b) cross section of one of the crotales.

Ratios of the principal modal frequencies to the fundamentals (2,0 mode) are given in Table 9.1 along with the corresponding modal frequency ratios in a flat plate. Note that the (3,0) mode, having three nodal diameters, sounds close to an octave above the fundamental, much as the case of the center-loaded gong, although here we are dealing with modes having nodal diameters rather than with axisymmetric modes. At any rate, the octave partial adds measurably to the sense of pitch.

Table 9.1. Frequency ratios for the principal modes of vibration in a set of tuned crotales.

Crotale	f_{01}/f_{20}	f_{30}/f_{20}	f_{11}/f_{20}	f_{40}/f_{20}	f_{50}/f_{20}
C_6	1.45	2.08	3.05	3.05	
$C_6^\sharp$		2.09	2.89	3.63	
D_6		2.01	2.99	3.44	
$D_6^\sharp$		2.13	3.38	3.66	
E_6	1.43	1.93	2.86	3.31	
F_6	1.39	2.01	2.81	3.51	
$F_6^\sharp$	1.42	1.98	2.85	3.41	5.19
G_6	1.42	1.97	2.81	3.38	5.17
$G_6^\sharp$	1.49	1.96	2.97	3.34	5.07
A_6		1.98	2.76	3.42	5.14
$A_6^\sharp$		2.00	2.88	3.41	5.15
B_6		1.94	2.71	3.29	4.97
C_7	1.42	1.92	2.67	3.28	4.92
Flat plate	1.73	2.33	3.91	4.11	6.30

9.10. Kyezee

The *kyezee* is a plate chime found in Burma, Indonesian, and Tibet. It is more or less triangular in shape with rather elaborate curved lines. A number of plates of differing size and pitch are suspended vertically and struck with a hard mallet. Modern kyezees or "Burma gongs" are often suspended from a pivot so that when struck they can spin around and give a rapidly modulated tone. Vibrational modes of a kyezee are shown in Fig. 9.19.

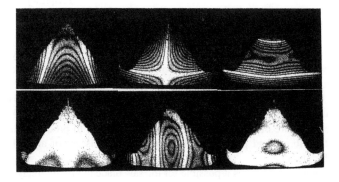

Fig. 9.19. Vibrational modes of a Burmese kyezee: 954, 1562, 1749, 2447, 4440, and 5349 Hz.

9.11. Bell Plates

Large metal bell plates, like chimes or tubular bells (see Section 7.5), are used in orchestras and bands to produce sounds with bell-like quality, as in Wagner's "Parsifal," for example. Various shapes and materials have been used, but some of the best results have been obtained with rectangular steel plates having length-to-width ratios around $L/W=\sqrt{2}$ (=1.41). Large steel bell plates are reported to have a more distinct strike note than tubular bells [18].

In constructing a set of tuned bell plates, it is convenient to make the thickness and the material of all the plates the same. Their lengths and widths will then be inversely proportional to the square root of the frequency. When a set of steel bell plates was fine-tuned by ear, the final length-to-width ratios that gave the best sound in the individual plates were found to range from 1.35 to 1.63, which includes the ratio $L/W=\sqrt{2}$ [18].

The sound spectrum of a steel bell plate tuned to A_4 (440 Hz) is shown in Fig. 9.19. This plate, with $L/W=1.40$, showed a hum note at 213 Hz, a fundamental at 438 Hz, and an octave partial at 875 Hz. These strong partials, due to the (2,0), (0,2), and (2,2) modes (See Section 8.4), are reasonably close to a harmonic series 1:2:4.

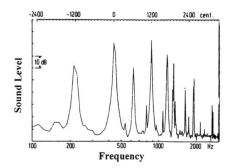

Fig. 9.20. Sound spectrum of a steel bell plate with $L/W=1.40$, tuned to A_4. Strong partials include a sub-octave (hum note), a fundamental, and an octave [18].

It is reported that large bronze bell plates sound about a quarter tone sharp on impact, gliding to their final pitch in about three seconds [1].

9.12. Musical Saw

The musical saw, a rather popular folk instrument, may be an ordinary carpenter's saw or a specially built saw of fine steel with or without saw teeth. The saw handle is held between the knees while the blade is bent into an S-curve and bowed along the edge with a cello or bass bow as shown in Fig. 9.19. This generally excites a single mode of vibration and results in a nearly sinusoidal tone. The player controls the pitch of the sound by changing the curvature of the blade, increasing the curvature to produce a higher pitch. Because of the vibrato and pitch glide, the sound is reminiscent of an early electronic instrument, the Theremin.

The principal that makes it possible for the saw blade to vibrate freely, irrespective of the damping action of the saw handle and the player's hand, is the internal reflection of transverse waves when the curvature exceeds a critical value. In a long, narrow plate, the

(2,0) mode and other modes of the (*m*,0) family (see Fig. 8.2) are simple bar-like vibrations, while modes of the (0,*n*) family (such as the (0,2) mode in Fig. 8.3) have *n* nodal lines parallel to the edge and are of high frequency when the plate is narrow.

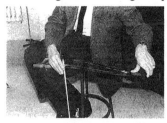 

Fig. 9.21. Chladni patterns (white powder) showing the (2,2) and (3,2) modes of vibration of a musical saw. Note that the handle is firmly held by the player's knees while the left hand applies the desired amount of stress by bending the blade to change the pitch. Shaking the leg provides vibrato (photographs courtesy of Arnold Tubis).

The modes of interest in the musical saw are the (*m*,2) nodes, which have 2 nodal lines running parallel to the long edges and *m* transverse nodes. These modes are excited by bowing the edge. (The (2,2) and (3,2) modes are shown in Fig. 8.3). It can be shown mathematically, and also demonstrated experimentally, that these $(m,2)$ modes are reflected when the blade curvature exceeds a certain critical value [19,20]. To raise the pitch, the player shortens the effective vibrating length of the blade by forcing the region of high curvature farther from the end of the saw.

9.13. Flexatone

Another instrument that depends upon a stressed plate is the flexatone, shown in Fig. 9.21. A thin flexible steel plate, fastened to the frame at one end, is hit on each side alternately by rubber or wooden beaters on springs. The pitch is changed by thumb pressure exerted on the end of the plate. Single strokes are possible, but the tremolo is more common.

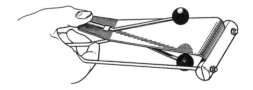

Fig. 9.22. Flexatone. The pitch of a steel plate is changed by exerting thumb pressure on the end of the plate.

The flexatone was introduced into jazz music in the 1920s, but has been used by composers such as Schoenberg, Honegger, and Henze. Probably the most famous passage is in Khachaturian's Piano Concerto, where the flexatone plays the melody line with the violins in the second movement [1].

References
1. J. Holland, *Percussion* (Macdonald and Jane's, London, 1978).
2. T. D. Rossing and C. Wilbur, *Proceedings, Intl. Symp. On Simulation, Visualization and Auralization for Acoust. Research and Education (ASVA97)* (Tokyo, 1997) 675.
3. E. F. F. Chladni, *Die Akustik* (Breitkopf and Härtel, Leipzig, 1830).
4. Lord Rayleigh, *The Theory of Sound*, Vol. I, 2nd ed. (Macmillan, London, 1894).
5. T. D. Rossing, *American J. Physics* **50** (1982) 271.
6. S. Schedin, P. O. Gren, and T. D. Rossing, *J. Acoust. Soc. Am.* **103** (1998) 1217.
7. T. D. Rossing and R. B. Shepherd, *Percussive Notes* **19**(3) (1982), 73.
8. N. H. Fletcher, R. Perrin and K. A. Legge, *Acoustics Australia* **18**(1) (1989) 9.
9. Fletcher, N. H., in D. Green and T. Bossomaier, eds, *Complex Systems: From Biology to Computation* (IOS Press, Amsterdam, 1993).
10. C. Wilbur and T. D. Rossing, *J. Acoust. Soc. Am.* **101** (1997) 3144.
11. C. Touzé, A. Chaigne, T. Rossing, and S. Schedin, in *Proceedings of ISMA98* (Acoustical Society of America, Woodbury, NY, 1998) 377.
12. J. Blades, *Percussion Instruments and Their History* (Faber and Faber, London, 1970).
13. T. D. Rossing and N. H. Fletcher, *Bull. Australian Acoust. Soc.* **10**(1) (1982) 21.
14. K. A. Legge and N. H. Fletcher, *J. Acoust. Soc. Am.* **86** (1989) 2439.
15. T. D. Rossing and R. B. Shepherd, *Percussive Notes* **19**(3) (1982) 73.
16. T. D. Rossing and N. H. Fletcher, *J. Acoust. Soc. Am.* **73** (1983) 345.
17. L. M. Nickerson, *The Acoustics of a Karen Bronze Drum*, MS Thesis, Northern Illinois University, 1999.
18. C.-R. Schad and G. Frik, *Acustica* **82** (1996) 158.
19. J. F. M. Scott and J. Woodhouse, *Phil. Trans. Roy. Soc. Lond.* **A339** (1992) 587.
20. N. H. Fletcher and T. D. Rossing, *The Physics of Musical Instruments*, 2nd ed. (Springer-Verlag, New York, 1998). 665.

Chapter 10
Music from Oil Drums: Caribbean Steelpans

The Caribbean steelpan is probably the most important new acoustical musical instrument to develop in the 20th century. In addition to being the foremost musical instrument in its home country, Trinidad and Tobago, steel bands are becoming increasingly popular in Europe, North America and some Asian countries as well. The modern family of steelpans now covers a 5-octave range, and steel bands of today use them to play calypso, popular, jazz, and Western classical music.

The development of steelpans took place in the years following the end of World War II, when the annual celebration of Carnival was resumed with great enthusiasm. Many claims have been made about the invention of the tuned steelpan. Undoubtedly, it resulted from a lot of trial and error on the part of musicians and inventors such as Bertie Marshall, Anthony Williams, and Ellie Mannette. Thousands of 55-gallon oil drums left on the beach by the British navy provided ample raw material for experimentation. Although the basic designs have pretty well stabilized, steelpans are still evolving.

Modern steel pans are known by various names, such as tenor (or lead), double second, double tenor, guitar, cello, quadrophonics, and bass. The overlapping ranges of these instruments are shown in Fig. 10.1.

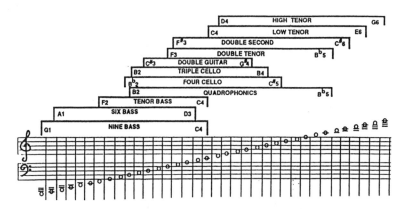

Fig. 10.1. Typical playing ranges of steelpans [1].

The design of a typical steelpan set is shown in Fig. 10.2. The tenor pan has from 26 to 32 different notes, but each bass pan has only 3 or 4; hence the bass drummer plays on six pans in the manner of a timpanist. Different makers still use different designs, although there is a movement in Trinidad and Tobago to standardize designs [2]. Most steelpans are played with a pair of short wood or aluminum sticks wrapped with strips of surgical rubber. The bass pans are hit with a beater consisting of a sponge rubber ball on a stick.

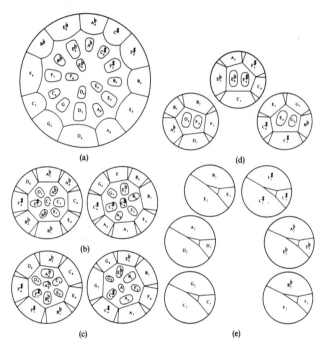

Fig. 10.2. Design of a typical steelpan set, showing locations of various notes: (a) single tenor; (b) double tenor; (c) double second; (d) cello; and (e) bass (courtesy of Clifford Alexis).

10.1. Construction and Tuning

The typical steelpan is constructed from a portion of a 55-gallon oil drum. The end of the drum is hammered ("sunk") into a shallow concave well, which forms the playing surface. The depth of the well varies from 10 cm (4 in) in a bass pan to 19 cm (7.5 in) in a tenor pan. Part or all of the cylindrical portion of the drum is retained as the skirt, which acts as a baffle, acoustically separating the top and bottom of the playing surface to prevent cancellation of radiated sound; it also radiates considerable sound, itself. Skirt lengths vary from about 13 cm (5 in) for the tenor to 45-70 cm in the cello pan. A bass pan retains the entire length of the original drum. Skirt lengths that give the best sound for each pan have been determined by trial and error.

After the end of the oil drum has been hammered into a shallow well, the playing surface is scribed and grooved with a punch to define the individual note areas, as shown in Fig. 10.2. The tuner raises each note area up slightly by hammering, and then works it up and down to soften the metal. Next the pan is heated over a bonfire, which raises the pitch of each note and appears to harden at least the surface of the metal. The note areas are flattened and then "tightened" around the edges by rapid hammer work known as "piening," which prepares the note areas for tuning.

The actual tuning process begins by lightly tapping the underside of the pan to create a small flat to convex portion of the note area. The grooves between the note areas on the playing surface are retightened with a small hammer, so that when the player strikes the note area, the vibrations will be restricted mainly to that note area. The fundamental frequency of a note area can be lowered by glancing blows across the top of the area, while it can be raised by increasing the height at the center or by tapping down at the corners of the area.

A skilled pan maker also tunes at least one overtone of each note area to a harmonic of the area's fundamental frequency. The first overtone will nearly always be the octave, and if the note area is large enough, a second overtone is tuned to the third or fourth harmonic. Tuning the third mode to the musical twelfth (third harmonic) gives the note a more mellow tone, tuning it to the double octave (fourth harmonic) gives it a bright tone. Tuning the harmonics is more difficult than tuning the fundamental frequency. In general, tapping the underside of the pan along or just outside a boundary of a note area that runs parallel to the nodal line for that mode will lower a harmonic, whereas tapping down or outward on the playing surface just inside this boundary will raise the harmonic frequency [3,4].

10.2. Normal Modes of Vibration

Like most other percussion instruments we have discussed, the vibrations of steelpans are conveniently described in terms of *normal modes of vibration.* A normal mode of vibration represents the motion of a structure at a normal frequency (eigenfrequency). A normal frequency is almost the same as a resonance frequency but not quite. A resonance, which represents maximum response of the structure when driven with an oscillating force, may be made up of a combination of normal modes having normal frequencies close to the driving frequency.

In theory, it should be possible to excite a normal mode of vibration at any point in a musical instrument that is not a node and to observe motion at any other point that is not a node. A normal mode is a characteristic only of the structure itself, independent of the way it is excited or observed. Normal mode shapes are unique for a structure, whereas the deflection of a structure at a particular frequency, called an operating deflection shape (ODS) may result from the excitation of more than one normal mode. Normal mode testing has traditionally been done using sinusoidal excitation, either mechanical or acoustical. Detection of motion may be accomplished by attaching small accelerometers, although optical methods are less obtrusive.

Normal modes of a complex structure, such as a steelpan, are functions of the entire structure. A normal mode shape should describe how every point on the pan moves when the instrument is excited at any point. As a practical matter, at any given frequency, some parts of the pan move much more than other parts. For example, we often speak of the normal modes of vibration of a particular note area, but we must always be aware of the fact that when one note area vibrates, all other parts of the pan vibrate at the same frequency, although their motion may be very small–too small to be observed, in many cases[5].

10.3. Interlude: Holographic Interferometry

Optical holography is a technique for the recording and reconstruction of three-dimensional images using photographic film or digital storage. The basic technique was developed by Dennis Gabor in 1947 primarily as a means for improving electron microscopy, and it eventually won him the Nobel prize in physics. Gabor's wavefront reconstruction process led to recorded patterns that he called holograms from the Greek word *holos*, meaning whole. His invention went more or less unused, however, until the invention of the laser in 1960, when Leith and Upatniek (United States) and Denisyuk (Russia) adapted the technique to produce optical holograms.

A hologram is a two-dimensional recording capable of producing a three-dimensional image. It does that by recording the whole wave field, including both the *phase* and *magnitude* of the light waves reflected from an object. The phase information is converted into intensity information (which can be recorded on film) by using a reference beam of coherent light.

The most striking feature of a hologram is the three-dimensional image that it forms. An observer looks through a hologram as if it were a window and sees a three-dimensional image on either side or even straddling this window. Moving the head from side to side or up and down allows the viewer to look around the object. Furthermore, if the hologram is broken into small pieces, the entire image can be seen through any piece, although the viewing window becomes much smaller. Holograms can provide either real or virtual images.

Making holograms requires a coherent light source, such as a laser. A common arrangement for producing transmission holograms is shown in Fig. 10.3 (a). Light from the laser is partially reflected and partially transmitted by the beam-splitting mirror (BS), so that two coherent beams result. The reference beam (R) goes directly to a photographic plate, while the object beam (O) illuminates the object. Scattered light from the object (S) also reaches the photographic plate, and the two beams form an interference pattern which records all the information needed to reproduce a 3-dimensional image when viewed through the processed plate in the reference beam, only, as in Fig. 10.3(b). For a more complete discussion, see [6].

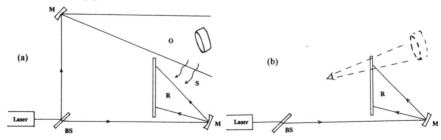

Fig. 10.3. (a) Arrangement for making a transmission hologram by splitting laser light into a reference beam (R) and an object beam (O); (b) Arrangement for viewing a hologram in the reference beam. If the plate is in the exact location it was for recording, the image will be in the location of the original object.

In addition to the primary fringes that appear in all holograms due to interference between the object and reference beams, much larger secondary fringes may appear if the object has moved during the exposure. These secondary fringes have been used to study stress, motion, and vibration. One way to record these fringes is by using the technique of time-average holographic interferometry.

A vibrating object is instantaneously stationary at the two extremes of its motion, while it moves at maximum speed somewhere near the center. On the average, therefore, it spends more time near the turning points, and so a holgram of it with a long exposure will show a pattern of interference fringes that indicates its displacement difference at the two turning points. This method was used to record the vibrational modes of a snare drum in Figs. 4.5 and 4.6, for example. The fringes provide a sort of contour map that indicates how much each point has moved between the turning points and therefore its amplitude of vibration.

Although holograms recorded on film generally have the highest possible resolution, the use of film is somewhat time consuming. It is much more convenient to create the holographic images electronically, so that they can be viewed as soon as they are created. In the 1970's video techniques to record holograms or speckle patterns, as they are sometimes called, were developed. TV holography or electronic speckle pattern interferometry allows real-time fringes to be presented on a TV monitor without any photographic processing.

Modern TV holography systems use sensitive CCD (charge-coupled device) cameras, and they incorporate image processing using digital computers and techniques such as phase stepping. TV holography has become popular in engineering laboratories and in industry to study vibrations, deformations, sound fields, and in nondestructive testing. One of its first applications in the arts has been the study of vibrations in musical instruments. It has not yet been widely used by visual artists to create real time holographic images, however.

An optical system for TV holography is shown in Fig. 10.4. A beam splitter BS divides the laser light to produce a reference and an object beam. The reference beam illuminates the CCD camera via a phase-stepping mirror PS and an optical fiber, while the object beam is reflected by mirror PM so that it illuminates the object to be studied. Reflected light from the object reaches the CCD camera, where it interferes with the reference beam to produce the holographic image. The speckle-averaging mechanism SAM in the object beam alters the illumination angle in small steps in order to reduce laser speckle noise in the interferograms. Most of the interferograms of the steel pan in this chapter were recorded using TV holography.

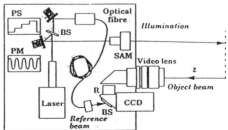

Fig.10.4. Optical system for TV holography. BS = beam splitter; PS = phase-stepping mirror; PM = mirror; SAM = speckle-averaging mechanism; CCD = video camera.

10.4. Modes of a Tenor Pan

The layout of the notes on a typical tenor pan is shown in Fig. 10.5. The outer ring has 12 more or less trapezoidal note areas tuned from D_4 (294 Hz) to $C_5^{\#}$ (554 Hz). The middle ring has 12 more or less elliptical note areas tuned from D_5 to $C_6^{\#}$ and the inner ring has four to six near circular note areas tuned from D_6 to $F_6^{\#}$. Note that the notes in the outer and middle rings are arranged in "circles of fifths": moving counterclockwise, one goes up a perfect fifth or down a perfect fourth (which is equivalent to going up a fifth and down an octave) on the musical scale. Each note in the middle ring is an octave higher than the corresponding note on the outside ring. The inside ring adds a third octave to 5 notes (6 notes when an optional $F_6^{\#}$ is included). Thus each note has several harmonics in common with its nearest neighbors, which leads to the strong interaction between notes characteristic of steel pans. (Another note arrangement, sometimes called the Invaders layout, is also used for tenor pans by some pan makers).

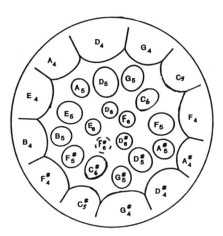

Fig. 10.5. Layout of note areas on a typical tenor pan.

Holographic interferograms of the tenor pan driven at several frequencies that excite modes of vibration in the C_5 note area (at two o'clock, viewing the pan as a clock face) are shown in Fig. 10.6. Note that several other note areas show appreciable vibration, especially at the higher frequencies. At the lowest frequency the active (roughly elliptical) portion of the note area moves in a single phase, while at the frequency of the second harmonic, there is a nodal line parallel to the rim dividing the note area into halves. At 1421 Hz ($2.72f_1$) there is a radial node; at 2064 Hz ($3.95f_1$) there are two nodal lines parallel to the rim, and at 2184 Hz, two nodal lines perpendicular to each other. Besides five modes of the C_5 note area, many sympathetic vibrations of other note areas are apparent in Fig. 10.6. The various note areas can be identified by comparing Fig. 10.6 with Fig. 10.5.

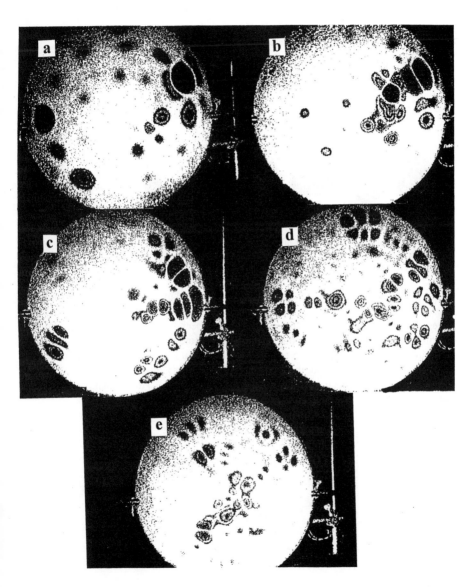

Fig. 10.6. Vibrations of the tenor pan at frequencies that excite modes in the C_5 note area: (a) 522 Hz; (b) 1050 Hz; (c) 1421 Hz; (d) 2064 Hz; (e) 2184 Hz. [5]

Holographic interferograms of several modes observed in the D_4, D_5, and D_6 note areas of the same tenor pan are shown in Fig. 10.7. These are designated by (m,n), where m is the number of circumferential nodal lines and n is the number of radial nodal lines.

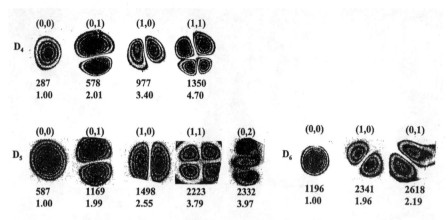

Fig. 10.7. Holographic interferograms of several modes in the D_4, D_5, and D_6 note areas of the tenor pan. Modes are designated by (m,n), where m and n are the numbers of circumferential and radial nodes, respectively. Frequency ratios to the fundamentals are given.[5]

Relative frequencies of the lowest modes observed in all 28 note areas are shown in Fig. 10.8. Note that the (0,1) is tuned to the second harmonic of the fundamental in all the notes. In the lowest six or seven notes, the (1,0) has been tuned to the third harmonic, with varying degrees of accuracy, although in the highest notes of the outer ring and throughout the middle ring it is closer to 2.5 times the fundamental frequency (an interval of a major tenth or an octave plus a major third above the fundamental). In a few notes, the (0,2) mode is tuned to the double octave, and in three notes ($F_5^\#$, $C_5^\#$, and E_5) the (1,1) mode fulfills that role.

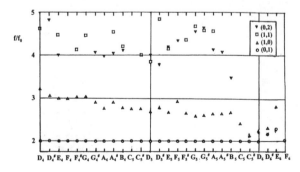

Fig. 10.8. Relative frequencies of the lowest modes in the Alexis tenor pan.[5]

10.5. Modes of a Double-Second Pan

The double-second steel pan plays in the alto range. It consists of two pans, each having about 14 to 16 notes. Thus the note areas are considerably larger than those of the tenor pan.

Modal shapes observed in three different B^b note areas in a 15-note double-second pan by Clifford Alexis are shown in Fig. 10.9 [7]. The lines in these sketches are nodal lines; regions on either side of a nodal line move in opposite directions. Because the note areas are larger in a double-second pan, it is possible to tune both second and third harmonics in all the note areas, and sometimes a fourth harmonic is tuned as well. In general the second- and third-harmonic modes are "orthogonal"; that is, the nodal lines are perpendicular to each other.

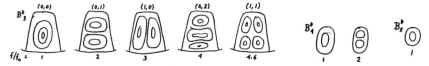

Fig. 10.9. Mode shapes in three different B^b note areas in a double-second steel pan [7].

Harmonic tunings of modes 2-5 in Fig. 10.9 are shown in Fig. 10.10 for all 15 notes in the drum. Note that the (1,0) mode is always tuned to the 2nd harmonic, and the (0,1) mode is generally tuned to the 3rd harmonic. However, either the (1,1) mode or the (2,0) mode may be tuned to the fourth harmonic (it is the (2,0) mode in the B_3^b note area). Tuning the (1,1) mode to the 4th harmonic is sometimes referred to pan tuners as "diagonal" tuning for obvious reasons.

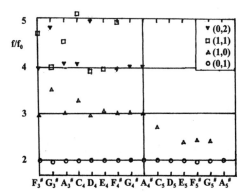

Fig. 10.10. Harmonic tuning of the (1,0), (0,1), (1,1) and (2,0) modes (modes 2-5 in Fig. 10.9) in a 15-note double-second pan by Clifford Alexis [7].

The double-second pan whose mode shapes are shown in Fig. 10.11 is a "single-grooved" pan; a single groove separates the larger note areas from each other. Most pan makers these days prefer a "double-grooved" design, such as the one shown in Fig. 10.2 (c). Mode shapes in a double-grooved double-second pan are shown in Fig. 10.11.

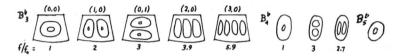

Fig. 10.11. Mode shapes in three different B^b note areas in an Alexis double-second pan with a "double-grooved" design.

Harmonic tunings of modes 2-5 in Fig. 10.11 are shown in Fig. 10.12 for all 14 notes in the pan. Note that the (1,0) mode is always tuned to the 2nd harmonic, and the (0,1) mode is generally tuned to the 3rd harmonic except in the two lowest notes where it is tuned to the 4th harmonic. In three notes the (1,1) mode is tuned close to the 4th harmonic, while in two notes (including B_3^b) it is the (2,0) mode that is tuned close to this harmonic.

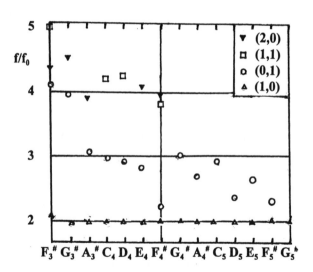

Fig. 10.12. Harmonic tuning of the (1,0), (0,1), (1,1) and (2,0) modes (modes 2-5 in Fig. 10.11) in a 14-note double-second pan by Clifford Alexis with "double-groove" design [7].

10.6. Sound Spectra

The sound spectra of steelpans are rich in harmonic overtones. These appear to have three different physical origins:
1. Radiation from higher modes of vibration of a given note area, tuned harmonically by the tuner;
2. Radiation from nearby notes whose frequencies are harmonically related to the struck note;
3. Nonsinusoidal motion of the note area, vibrating at its fundamental frequency.

Sound spectra of two notes on a double second steel pan are shown in Fig. 10.13. Note that harmonics as high as the ninth harmonic are detected. The first three or four harmonics result from higher modes harmonically tuned, while the radiation from harmonically-related nearby notes excited by sympathetic vibration could contribute to harmonics such as two, four, six, and possibly eight. The balance of the harmonics are attributed to non-sinusoidal motion of the note area.

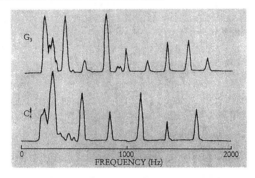

Fig. 10.13. Sound spectra from two notes on a double-second steelpan [8].

Nonsinusoidal Motion of the Note Area

Non-sinusoidal vibrational motion results from asymmetry in the stiffness of a note area due to its shape. Pressing down on an arch, for example, results in less deflection than pushing upward. The note areas in a steelpan, although nearly flat, have rather complicated shapes that include both upward and downward bubbles or "arches" in local areas. In studying this effect several years ago, we found that when a note area is driven at its fundamental frequency, the amplitude of the second harmonic increased as the square of the fundamental amplitude, the third harmonic as the cube of the fundamental amplitude, and the fourth harmonic as the fourth power of the fundamental amplitude [9]. Beyond the fourth harmonic, it was difficult to establish a quantitative relationship between fundamental amplitude and harmonic amplitude, but harmonics out to the eighth were observed at large fundamental amplitude. These measurements, incidentally, were made with the other note areas damped with sandbags so that sympathetic vibrations of harmonically-tuned notes could be discounted.

Mechanical Coupling Between Note Areas

Striking a note area in a steelpan excites the note areas that are tuned to a harmonic of the struck area. For example, striking the B_3^b note area excites the B_4^b note (2nd harmonic), the F_5 note (3rd harmonic), the B_5^b note (4th harmonic), etc. The amplitude of sympathetic excitation of the harmonic note is a nonlinear function of the amplitude of the fundamental note; that is, doubling the amplitude of the fundamental will more than double the amplitude of the harmonics, so a graph of harmonic amplitude *vs* fundamental amplitude is not a straight line (non-linear).

Figure 10.14 shows the vibrations in a double second pan when the B_3^b note area is driven sinusoidally at its fundamental, second harmonic, and third harmonic frequencies. The force amplitude increases from top to bottom. Note that driving the B_3^b note area at its second harmonic frequency (center column) not only excites the second mode in the B_3^b note area but also the fundamental in the B_4^b note area just inside it. Similarly, driving the B_3^b note area at its third harmonic (right hand column) excites the 3rd mode and also weakly drives the E_5 note, which is a semi-tone lower, off resonance (the F_5 note, which would be resonantly excited is on the other pan).

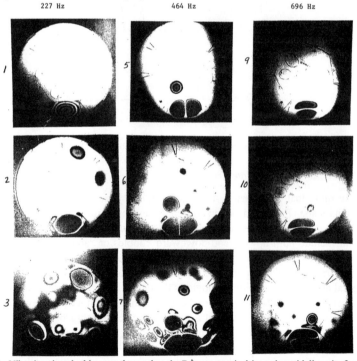

Fig. 10.14. Vibrations in a double second pan when the B_3^b note area is driven sinusoidally at its fundamental, second harmonic, and third harmonic frequencies. The force amplitude increases from top to bottom.

Figure 10.15 shows the B_4^b note area in the same double-second pan driven at its fundamental, second harmonic, and third harmonic frequencies. Note that the B_3^b note area now vibrates in its (1,0) mode (first column) and (2,0) mode (second column). At 1251 Hz the A_4^b note (at one o'clock) is driven in its (1,0) (third) mode which is three times the fundamental frequency. As the amplitude increases, many other note areas participate as well.

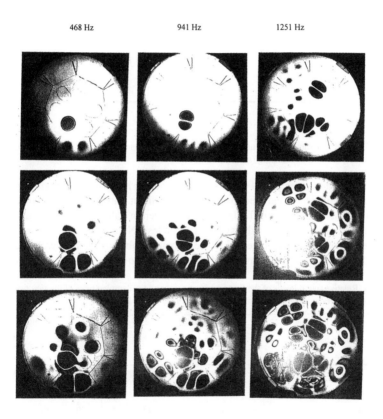

Fig. 10.15. Vibrations in a double second pan when the B_4^b note area is driven sinusoidally at its fundamental, second harmonic, and third harmonic frequencies. The force amplitude increases from top to bottom.

The sound spectrum of the A_3^b note as ordinarily played (approximately *mf*) and the spectrum of the same note with the sympathetically resonant A_4^b and A_5^b note areas damped by sand bags are compared in Fig. 10.16. Note that the harmonic peaks without damping are 5 to 10 dB higher than with damping of sympathetically resonant notes.

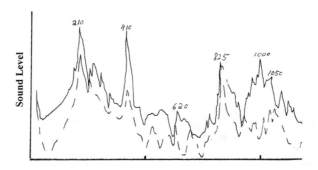

Fig. 10.16. Sound spectrum of the A_3^b note of a double second steel pan: (a) with the A_4^b and A_5^b notes damped (dashed curve); (b) with no damping (solid curve) [10].

The relative strengths of the first four harmonics in the the seven notes in the outer circle of the double second steel pan with and without damping of the sympathetically resonant notes are compared in Table 10.1. Also given in Table 10.1 are the damping rates with and without damping of the sympathetic notes. Note that the damping rate of the fundamental note is slightly increased by the damping of the other notes, while the damping rate of the second harmonic is increased by a greater amount.

Table 10.1. Relative sound levels of the first four harmonics and the damping rates of the first two harmonics with and without damping of the sympathetically resonant notes [10]

Note	Relative harmonic level Without damping				Relative harmonic level With damping				Damping rate Without		With	
	$n=1$	2	3	4	$n=1$	2	3	4	$n=1$	2	$n=1$	2
$F_3^\#$	0	-8	-18	0	0	-10		-1	50 dB/s	50	80	40
$G_3^\#$	0	-1	-23	3	0	-10	-21	-10	20	20	30	70
B_3^B	0	-22	-10	-9	0	-19	-11	-7	80	20	140	120
C_4	0	-15	-20	-9	0	-8	-22	-9	10	30	50	70
D_4	0	-12		-17	0	-20	-23	-27	120	10	140	10
E_4	0	-7	-25	-9	0	-14	-26	-7	50	130	70	70
$F_4^\#$	0	-11	-26	-6	0	-15	-31	-7	60	50	40	110

10.7. Note Shapes

The note areas in the various steelpans differ widely in size and shape. Some are nearly trapezoidal, some are oval, and some are nearly circular. When struck with a soft beater, a note area will vibrate in a complex manner which can be analyzed in terms of normal modes of vibration. The pan tuner has many options at his/her disposal for these modes of vibration to harmonics of the fundamental. The (0,1) mode is nearly always tuned to the second harmonic, and in many notes, the (1,0) mode is tuned to the third harmonic (a musical twelfth above the fundamental) or to a frequency about 2.5 times the fundamental (a major tenth above the fundamental). Although most tuners use electronic tuners, considerable reliance is placed on their experienced ears. The exact tuning is determined partly by how a particular note responds; when it is not possible to tune modes to the third or fourth harmonics, for example, the tuner adjusts the note to give the "best" sound.

The rather daunting task of designing the layout of notes on various pans has been largely a matter of trial and error. Although tuners still customize pans for certain players, there has been a movement toward standardizing note layout [2]. Likewise, there has been a trend toward standardization of note size and shape. We have found that the longest dimension L of note areas in a large variety of pans appears to follow a scaling law $L=Kf^{-2/3}$, where f is the fundamental frequency [3]. There is little known scientific basis for this empirical formula, however, and it may or may not be optimum.

The only known attempt to model note areas is that of Hampton [7], who modeled them as shallow rectangular shells with length L, width W, thickness h, and radii of curvature R_L and R_W along the length and width, respectively. Using a computer, Hampton tested a total of 143,520 physical configurations for various mode hierarchies. About 1.2% of the configurations tested produced a second harmonic (1,0) mode. About 4% of those that produced a second harmonic also produced a third-harmonic (0,1) mode; about 9% of them produced a fourth-harmonic (2,0) mode. In most of the harmonic cases, the profile parameter $R=R_W/R_L$ was negative and the curvature parameter R_W/W was small (less than 10).

Profiles of note areas in pans tuned by various tuners reveal the rather surprising result that most of them are nearly flat. Although the notes appear convex, when viewed against the concave surface of the pan, they are, in fact nearly flat, with local areas of both positive and negative curvature. The mode frequencies of a flat rectangular plate with simply-supported (hinged) edges are fairly easy to calculate. For a plate with $L/W=\sqrt{2}$, the lowest three modes are harmonically related, and the (1,1) mode has 4 times the frequency of the fundamental for all values of L/W. The boundary condition at the edge of a note area might be expected to be somewhere between simply-supported and clamped, however.

Although a number of scientific investigations of steelpans have taken place [3-15], scientific studies on these instruments have not kept pace with their growing popularity. Results of acoustical studies are particularly valuable to pan designers who wish to be guided by scientific principles in improving the sound of pans. Tuning notes with a hammer presumably changes the local thickness and the local stresses. The effect of thickness change on a given mode of vibration should be most effective in an area of greatest deflection curvature which will ordinarily be at an antinode (the "bullseye" in the holographic interferogram of that mode). Thinning a plate reduces the mass but it also reduces the stiffness and the latter effect dominates (see Chapter 8), so modal frequencies are decreased by thinning. Glancing blows in which the hammer has a large component of motion parallel to the pan surface are used to change internal stress; the effect is sometimes described as "tightening" a note. Murr, et al. [14] found that hammering reduces grain size and that harmonic tuning may involve a complex interaction between plastic and elastic properties of the metal.

10.8. Metallurgy and Heat Treatment

Steelpans were originally constructed from used oil drums. The pan maker had little control over the quality of the raw material, a constant source of frustration. Many pan makers now purchase steel barrels directly from the manufacturer, and specify standards of quality. Still, steel barrels vary widely in quality, and it is difficult for the pan maker to test them and to avoid investing many hours in shaping an inferior barrel which will result in a steel pan of disappointing quality.

Oil drums are usually fabricated from low-carbon steel (0.01 to 0.10 C). In one drum, Murr, et al. [14] found a composition of 0.01% carbon, 0.03% phosphorus, 0.11% manganese, 0.02% chromium, and less than 0.01% each of silicon, molybdenum, and zinc. The drum skirt composition was 0.03% carbon, 0.03% phosphorus, 0.3% manganese, 0.03% chromium, and less than 0.01% each of silicon, molybdenum, and zinc. They are characterized by "fairly dense tangles and poorly formed dislocation cells with dislocation densities ranging from 10^8 to 10^9 cm^{-2}" [14]. Ferreyra, et al. [15] conclude that) 0.04-0.05% carbon is optimum for strain-ageing effects to produce an adequate level of hardening.

Sinking the drum by sledgehammer blows changes its metallurgical structure. Ferreyra, et al.[15] found that the Vickers hardness of hammered drums increased from 113 in the undeformed material to 170 after sinking. Heat treatment of the drum increases the hardness by another 5% or more due to strain ageing, the amount of change depending upon the carbon content of the steel among other things. Modal frequencies in the note areas increase 10 to 30% after firing [3].

Rohner, on the other hand, presses the drumhead to shape and then increases the surface hardness of the material by "sandwich hardening" it in a hot bath of nitric acid which increases the surface hardness while maintaining a softer interior, a property that makes tuning easier but maintains durability and tuning stability. He has measured the Vickers surface hardness to be 700 or more.

Thickness of the metal

In addition to changing the metallurgical structure and hardness, sinking the drum by hammering stretches and thins the metal nonuniformly, making it thinner at the center and thicker near the rim. In a pan constructed by Ellie Mannette, Murr et al. [14] found the thickness at the center to be about one-half of the thickness at the edge. On the other hand, in pans by Clifford Alexis, we find the thickness at the center to be about 70% of the thickness at the edge. No doubt this reflects a difference in the way the pan is sunk. In Rohner's pans, the playing surface is cold-pressed into shape rather than hammered, and the thickness variation is less than 10%. The center notes are thicker than those of the hammered pans, the outer notes are thinner, and the notes are uniform in thickness.

Heat Treatment

Originally, the fully patterned and grooved steelpan was placed over an open fire to burn away the residual oil and paint. As the instrument evolved, this firing was recognized to have a significant effect on the sound of the finished pan, and closer attention was paid to the technique of firing. Typical firing temperatures are 300 to 400°F (150-200°C), although the pan maker is guided mostly by color changes in the heated drum head [15]. The drum is water cooled (quenched) or air cooled.

Ferreyra, et al. [15] found the hardness change during heat treatment to depend upon the carbon content of the steel and also on the amount of stress introduced by hammering. Heating to 350°F (177°C) for 10 minutes, followed by air cooling produced the most uniform hardness over both the center and outer notes. Higher temperatures, which produced a slightly greater hardness in the center notes, left the outer notes considerably softer. Thus, they concluded that the relatively low heat treatment temperatures used by most pan makers are probably near the optimum.

Why heat treatment raises the frequencies of the note areas [3] remains somewhat of a mystery. Probably it is due to relief of compressive stresses that result from sinking the pan. Murr, et al. [14] have suggested that frequency variations during heat treatment probably result from shape changes in the notes caused by elastic-plastic interactions. Changes in timbre during heat treatment have not yet been studied, as far as I know.

10.9. Skirts

The skirt of a steel pan serves as an acoustic baffle which enhances the efficiency of sound radiation, much the same way as a baffle enhances the radiation efficiency of a loudspeaker. The skirt length should be a substantial fraction of an acoustic wavelength. The high frequency notes of a tenor thus require only a small baffle, while the lower drums require successively longer skirts and the bass pan air cavity generally remains fully enclosed. Skirt lengths vary from about 13 cm (5 in) for the tenor to 45-70 cm in the cello pan. There has been a trend in recent years toward longer skirts, however.

Skirt vibrations and sound radiation

When a steelpan is played its skirt vibrates and radiates sound. Although measuring the total amount of sound radiated by the skirt is difficult, we have compared the vibrational spectra and amplitudes of the skirt with those of the struck notes. Using accelerometers attached to the pans, we found that in a tenor pan the vibrational amplitude of the skirt on the side nearest the struck note averaged 19±9 dB less than the note amplitude (averaging over all the notes), while it was 23±11 dB less than the average note amplitude on the side of the drum opposite the note. In a double-second pan, the corresponding differences in amplitude averaged 27±12 dB on the same side and 37±14 dB on the opposite side. In a cello pan the differences were 41±10 dB when the skirt accelerometer was on the same side as the note and 46±13 with it on the opposite side; in the bass pan it was 40±8 dB on the same side and 41±8 on the opposite side [3].

Whether the note was played loudly or softly seemed to make no significant difference in the above data. Even though the area of the skirt is considerably greater than that of the note area, the large differences in acceleration level (corresponding to vibration amplitude ratios ranging from about 10:1 to 200:1) suggest that radiation from the skirt does not contribute a lot to the sound level. However, the vibrational spectrum of the skirt is sufficiently different from that of the note area to influence the timbre of the pan.

Modal shapes in the skirt of the Alexis tenor pan are shown in Fig. 10.17. The modes correspond to standing bending waves propagating around the ring. At very low frequency (below 200 Hz), the modes have a nodal line at the center and antinodes at the top and bottom edge, as shown in Fig. 10.17(a), in spite of the rim at the top. From 200 Hz to about 1000 Hz, the modes have a node at the top rim and n antinodes along the bottom edge of the skirt, as shown in Fig. 10.17(b,c). Above 1000 Hz, the modes have a second nodal ring half a wavelength above the lower edge, as shown in Fig. 10.17(d) [5].

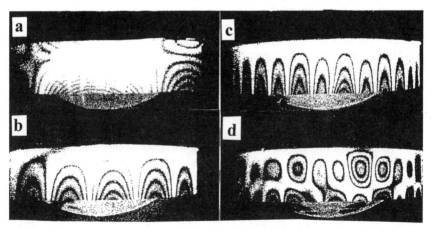

Fig. 10.17. Vibration modes of the skirt on the Alexis tenor pan: (a) 107 Hz; (b) 294 Hz; (c) 667 Hz; (d) 1400 Hz [5].

From holographic interferograms of the skirt, such as those shown in Fig. 10.17, it is possible to determine the wavelength and thus the speed of the bending waves. A graph of wave speed vs the square root of frequency is shown in Fig. 10.18. The bending wave speed in a thin plate is given by: $v=(1.8fhc_L)^{1/2}$, where $c_L=(E/\rho[1-v^2])^{1/2}$ is the speed of longitudinal waves (Fletcher and Rossing, 1998). For a steel plate with a thickness $h=0.9$ mm and $c_L=5050$ m/s, this predicts $v=2.86f^{1/2}$. The slope of the regression line in Fig.10.18 is 2.54 m/s, which is reasonably close to the value predicted by the simple theory for a flat steel plate of the same thickness.

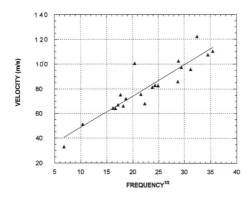

Fig. 10.18. Bending wave velocity in the skirt of the Alexis tenor pan (calculated from modes of vibration such as those in Fig. 10.17) [5].

10.10. Pans of Other Sizes

Steelpans made from 55-gallon oil drums are 22½ inches in diamter. Several pan makers have experimented with pans of a larger diameter. Panyard, Inc. offers an extended range lead pan 25½ inches in diameter with 33 notes from A_3 to F_6. Panart in Switzerland offers a family of specially hardened Blackpan instruments 600 mm (23.6 inches) in diameter.

Smaller pans are also made. Panyard, for example, offers mini-pans of 14-inch and 15¼-inch diameters. These smaller pans are easily carried in parades by means of a neck strap.

10.11. Recent and Future Developments

The Caribbean steelpan is a folk instrument that has found its way into the world's concert halls. This is partly due to the skill of young performers, now trained in universities and music conservatories, and partly due to the skill of craftsmen who have developed the fine art of pan making. Like makers of fine violins, these pan makers have learned their art by observing and serving as apprentices to other pan makers and by considerable trial and error. Crafting fine instruments, however, takes a great deal of time and skill, and as the demand for quality instruments begins to exceed the supply, fine instruments are becoming expensive. As in the field of violin making, the questions arises as to how less experienced makers, guided by scientific testing and measurements, can build and tune fine instruments. This appears to be possible as the acoustics of steel pans becomes better understood through scientific research.

Steelpans were originally constructed from used oil drums. The pan maker had little control over the quality of the raw material, which was a constant source of frustration. Many pan makers now purchase steel barrels directly from the manufacturer, and specify standards of quality. Still, steel barrels vary widely in quality, and it is difficult for the pan maker to test them and to avoid investing many hours in shaping an inferior barrel which will result in a steel pan of disappointing quality. Recent studies of metallurgy [14] and heat treating [15] of steel pan materials are a step in the right direction. We have made measurements on a steel pan constructed from custom-made blanks that include a tuning surface cold-pressed (rather than hammered) into shape and sandwich hardened by chemical treatment [5]. It seems likely that other pan makers will experiment with such materials, and that careful studies of their mechanical (such as in ref. 16) and metallurgical properties will continue.

The importance of materials in musical instruments, including the steel pan, is being recognized. In the past four years, the Materials Research Society, in cooperation with the Musical Acoustics Technical Committee of the Acoustical Society of America, has sponsored three special symposia on Materials in Musical Instruments (papers from the first symposium were published in the March 1995 issue of *MRS Bulletin*). A chapter on "Materials for Musical Instruments" has been included in the 2nd edition of *The Physics of Musical Instruments* [17].

The effect of skirt size on both the total sound radiation and the timbre should be determined. In custom blanks, the thickness of the skirt can also be varied to determine the effect on timbre.

At the Swiss company Panart, Felix Rohner and his associates, who have pioneered the use of custom sandwich-hardened material, have developed a new design which they call a *ping*. At the center of each note area is a small raised dome that provides a sturdy and stable strike point for the player and is designed to improve the sound as well.

Several pan makers in Trinidad have experimented with "bored" pans in which a series of bored holes separate the note areas. The one bored pan that we have studied in our laboratory does not appear to behave much differently from conventional "grooved" pans.

Although tuning techniques have developed mainly by trial and error, a scientific understanding of the pan should, in the future, enable tuning to proceed with greater accuracy and less effort. Given its growing popularity, it is safe to say that Caribbean steelpans will continue to develop as musical instruments.

References

1. L. Pichary, *Results of the Steel Pan Survey 90* (Trinidad and Tobago Bureau of Standards, Tunapuna, 1990).

2. K. Roach, *The Imperatives for Standardization* (Trinidad and Tobago Bureau of Standards, Tunapuna, 1992).
3. T. D. Rossing, D. S. Hampton, and U. J. Hansen, *Physics Today* **49**(3) (1996), 24.
4. U. J. Hansen, T. D. Rossing, E. Mannette, K. George, *MRS Bull.* **20**(3) (1995), 44.
5. T. D. Rossing and U. J. Hansen, *J. Acoust. Soc. Am.* (to be published)
6. T. D. Rossing and C. J. Chiaverina, *Light Science*. (Springer-Verlag, New York, 1999). Chapter 9.
7. D. S. Hampton (1995). *Investigation of the Vibrational Modes of Steel Drums by Holographic Interferometry* (MS thesis, Northern Illinois University, DeKalb).
8. T. D. Rossing, D. S. Hampton, and J. Boverman, *J. Acoust. Soc. Am.* **80**, S102 (abstract).
9. T. D. Rossing and J. Defrance, paper at Illinois Academy of Science, Chicago, 1991.
10. K. K. Leung and T. D. Rossing, paper at joint annual meeting of American Physical Society and American Association of Physics Teachers, Washington, DC, 1987.
11. A. Achong, *J. Sound Vibration* **191**, 207-217 (1996).
12. A. Achong, *J. Sound Vibration* **197**, 471-487 (1996).
13. A. Achong and K. A. Sinanan-Singh, *J. Sound Vibration* **203**, 547-561 (1997).
14. L.E. Murr, E. Ferreyra, J. G. Maldonado, E.A. Trillo, S. Pappu, C. Kennedy, J. DeAlba, M. Posada, D. P. Russell, and J. L. White, *J. Materials Science* **34**, 967-979.
15 .E. Ferreyra, J. G. Maldonado, L. E. Murr, S. Pappu, E. A. Trillo, C. Kennedy, M. Posada, J. De Alba, R. Chitre, and D. P. Russell, *J. Materials Science* **34**, 981-996.
16. A. Achong, *J. Sound Vibration* **212**, 623-625 (1998).
17. N. H. Fletcher and T. D. Rossing, *The Physics of Musical Instruments*, 2nd ed. (Springer-Verlag, New York, 1998).

Chapter 11
Church Bells and Carillons

Bells have been a part of almost every culture in human history. They are one of the most cherished musical instruments. Bells existed in the Near East before 1000 B.C., and in China at least as early as 1600 B.C. Only recently have modern scientific instruments allowed us to understand many of the subtleties of bell sounds. In this chapter we consider the science of casting, tuning, and playing Western church bells and carillon bells. In the following chapters, handbells and Eastern bells will be considered.

Although they are distinctly different bell types, tuned church bells (swinging) and carillon bells (stationary) bear a close resemblance to each other, and historically they developed in parallel. Although casting and tuning practices vary somewhat from country to country, there is a close resemblance between, say, English bells, Dutch bells, and French bells.

Bells developed as Western musical instruments in the seventeenth century when bell founders discovered how to tune their partials harmonically. The founders in the Low Countries, especially the Hemony brothers (François and Pieter), took the lead in tuning bells, and many of their fine bells are found in carillons today. Their understanding of the acoustics of bells was guided, to a large extent, by the research of carillonneur Jacob van Eyck, a distant relative of astronomer and mathematician Christiaan Huygens. Van Eyck concluded that the best sounding bells had five partials tuned harmonically to form intervals of an octave, a minor third, and a fifth with respect to the strike note (e.g., C_1, C_2, E_2^b, G_2, and C_3) [1].

After the deaths of the Hemony brothers, however, the art of founding and tuning fine bells declined, and was not rediscovered until the late 19th century by Canon Arthur Simpson in England. Simpson studied English bells as well as bells from Belgium and France, and concluded that none of these had a pure inner tonal structure. Guided in part by the scientific studies of Lord Rayleigh [2], he proceeded with systematic research to find out how to tune a bell on a lathe [3]. The Taylor foundry in Loughborough followed Simpson's recommendations in tuning bells. The Taylor carillon in Appingedam (1911) established Simpson's tuning methods in the Netherlands, and it was adopted by Dutch foundries as well.

Many fine bells were casualties of World War II. Germany had no natural source of copper and tin, both of which are essential to modern warfare. Hitler consequently ordered carillon bells in Holland and Belgium, as well as in Germany itself, to be shipped to Hamburg to be melted down as necessary. At the end of the war, hundreds of these bells still remained in Hamburg, and a large operation was undertaken to identify them and return them to their original towers.

Never before had so many bells of all shapes and sizes been collected in one place, and it was a golden opportunity for study and analysis. Financed by the Dutch government, E. W. Van Heuven set about analyzing their shapes and harmonic characteristics and was able to determine the effect of shape on the tuning characteristic which he included in his PhD thesis [4].

A typical profile of a tuned church bell is shown in Fig. 11.1. Note the thick soundbow where the metal clapper strikes. Most modern bells are cast of a bronze with a composition of approximately 80% copper and 20% tin. The inside of a bell is often turned down slightly on a lathe at various heights as individual partials are tuned, but the outside is left pretty much as cast.

11.1. The Carillon

Carillons are the world's largest musical instruments, by weight, at least. They have been playfully described as a means for making the "original heavy-metal music." A carillon consists of 23 to 77 tuned bells, a keyboard or clavier, and suitable links between the keyboard and the bells.

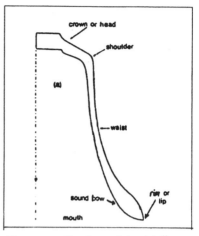

Fig. 11.1. Profile of a tuned church bell or carillon bell.

The largest carillon in the world is the Laura Spellman Rockefeller carillon in the Riverside Church in New York, shown in Fig. 11.2. A gift of the late John D. Rockefeller, Jr., it occupies a chamber near the top of a 392-foot tower. It includes 74 bells, from the 20-ton Bourdon to the smallest treble bell of 10 pounds; the total weight of all the bells is over 100 tons! Only slightly smaller is the Laura Spellman Rockefeller carillon in Chicago, a gift from the same donor. The Bourdon in the Chicago carillon, with a diameter of 117 inches, is tuned to $C_2^{\#}$, a semi-tone higher than the C_2 of its New York namesake.

Fig. 11.2. The Laura Spellman Rockefeller carillon at The Riverside Church in New York, with a total weight over 100 tons, is the world's largest carillon and the largest musical instrument in the world.

Although the playing of tunes by chiming bells dates back to the Middle Ages, the carillon probably came into being around 1500. In 1510, Jan van Spiere in the town of Oudenaarde mounted clappers in a set of chime bells which could be played with the aid of a keyboard. By 1530, the bell-founding families Waghevens and Van den Ghein in the Flemish city of Mechelen had brought the art of carillon building to the level at which it was to remain until the end of the 16th century [1]. Then came the Hemonys.

François Hemony was born in 1609 and his brother Pieter in 1619 at Levécourt in the Lorraine region of France, where many bell founders lived. They made frequent trips to the Low Countries and Germany to cast bells and other bronze articles, and during the Thirty Year War (1618-1648) they left Lorraine for good. In 1643, François Hemony was commissioned by the town of Zutphen to cast a carillon, which brought him into contact with Jacob van Eyck, an eminent musician who was advisor to the town. The three of them developed the art of bell tuning to new heights. In Zutphen the Hemony brothers cast fourteen fine carillons for various cities in The Netherlands before they moved to Amsterdam in 1657 [1].

Van Eyck and the Hemonys discovered that if the lowest overtones of the bell constituted a chord, the bell would have a pleasing sound and an unambiguous pitch. What is more important, they discovered the bell profile that would produce the ideal tuning. After casting, they did the final tuning by thinning the wall in various places on a bell lathe, a tuning practice that is still followed today.

The keyboard or clavier from which a carillon is played is not unlike a piano keyboard, but instead of flat keys it generally has round batons that are pushed down by the fist of the carillonneur as shown in Fig. 11.3. Activating the heavy clappers of the larger bells, even when they are properly balanced, requires appreciable force, and the carillonneur often wears a leather patch on his/her fist. The largest bells are activated by pedals as well.

Fig. 11.3. Carillonneur Margo Halsted at the keyboard of the Baird carillon at the University of Michigan.

Various mechanisms are used to transmit the playing force from the keyboard to the clappers. One of the oldest mechanisms, the breech wire system, is shown in Fig. 11.4(a). A horizontal wire is stretched between the clapper and a fixed point in the bell chamber, so that pressing the key pulls down the vertical wire and swings the clapper against the bell. A slightly more complicated mechanism is shown in Fig. 11.4(b). Obviously it is important to keep the playing mechanism in optimal adjustment, and before playing the bells, the carillonneur generally checks and adjusts the connecting wires, since temperature changes affect the mechanical linkages.

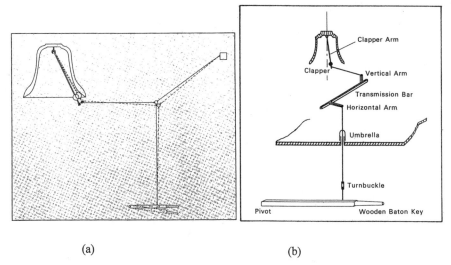

Fig. 11.4. Two examples of mechanical linkages between a carillon key and a bell clapper. (a) breech wire mechanism; (b) mechanism with a transmission bar.

The first carillon in North America, a set of 23 bells cast at the Bollée foundry in Le Mans, France and installed at Notre Dame University in 1856, originally had neither keyboard nor clappers. Fulfilling the dream of Father Edward Sorin, who attended a seminary in Le Mans, the bells were played by outside hammers actuated by an automatic cylinder mechanism or by players that pulled on knobs at the ends of handles. A traditional baton clavier was not installed until the mid 1950s, a hundred years later [5].

Most carillons are placed high in bell towers, as shown in Fig. 11.5, where they can be heard for great distances. Sometimes the bells are entirely open to the elements; more often, they are enclosed in a bell chamber with louvered windows. Very little research has been done on the acoustics of bell chambers, and the design of the bell chamber is often determined on the basis of tower appearance rather than on acoustical grounds.

A good argument can be made for having a hard reflecting ceiling above the bells in order to direct the upward radiated sound outward and downward, as shown in Fig. 11.6(a). Many chambers have a floor as well, so the sound undergoes several reflections, as shown in Fig. 11.6(b), inevitably losing 5 to 10% of the sound energy each time. Sometimes the bells are placed in more than one chamber, and a partial enclosure with louvres offers an opportunity to balance the sound of high and low bells, for example, as in Fig. 11.6(c). Each partial in the bell sound has a different radiation pattern, so if all the sound is radiated directly without reflections, a slightly different timbre will be heard in each direction. Atmospheric refraction and reflections from the ground and nearby buildings provide some mixing, however, even if the tower is completely open.

Fig. 11.5. Bell towers.

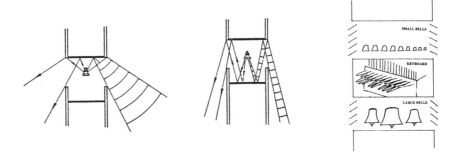

Fig. 11.6. Examples of bell chamber design. (a) a hard reflecting ceiling reflects sound outward and downward; (b) multiple reflections from ceiling and floor; (c) two bell chambers with louvres to balance the sounds.

11.2. Vibrational Modes of Church Bells and Carillon Bells

When struck by its clapper, a bell vibrates in a complex way. In principle, its vibrational motion can be described in terms of a combination of the normal modes of vibration whose initial amplitudes are determined by the distortion of the bell when struck. It is customary to classify these modes into families or groups with some common property of the nodal pattern. The most important families are those that have an antinode where the clapper strikes the bell just above the soundbow. In general, each normal mode of vibration contributes one *partial* to the sound of the bell. These partials are customarily given names such as *hum, prime, minor third* (or *tierce*), *fifth* (or *quint*), *octave* (or *nominal*), *upper octave*, etc. The strike note of the bell, which is determined by three partials (the upper octave, the upper fifth, and the octave), is generally close to the pitch of the prime in a well-tuned bell.

The first five modes of a church bell or carillon bell are shown in Fig. 11.7. Dashed lines indicate the locations of the nodes. The numbers (m,n) at the top denote the numbers of complete nodal meridians extending over the top of the bell (half the number of nodes observed along a circumference), and the numbers of nodal circles, respectively. Note that there are two modes with $m=3$ and $n=1$, one with a circular node at the waist and one with a node near the soundbow. Thus, we follow the suggestion of Tyzzer [5] and others and denote the one as $(3,1^{\#})$ in Fig. 11.7. The ratio of each modal frequency to that of the prime is given at the bottom of each diagram.

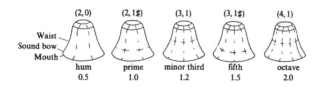

Fig. 11.7. The first five vibrational modes of a tuned church bell or carillon bell. Dashed lines indicate the nodes. Names of the partials and the frequency ratio to the prime are given below each diagram [6].

Most of the important modes of vibration in a bell are described as *inextensional*, in the sense that a neutral circle in each plane normal to the bell's symmetry axis remains unstretched. The radial and tangential components of the motion, u and v, respectively, are related by $u+\partial v/\partial\theta=0$, where θ is the polar angle the plane concerned [2]. Thus, we may write $u=m \sin m\theta$ and $v=\cos m\theta$. Motion of the bell for $m=0, 1, 2$, and 3 is illustrated in Fig. 11.8. For $m=0$, the motion is purely tangential, and we can describe these modes as "twisting" modes. Modes with $m=1$ might be described as "swinging" modes. As m increases, these modes have radial components that become increasingly larger compared with their tangential ones, and it is these radial components that are responsible for radiating sound.

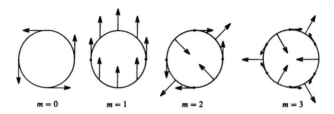

Fig. 11.8. Motion of a bell for inextensional modes of small m. Modes with $m=0$ and $m=1$ require one or more nodal circles ($n>0$).

A periodic table showing some of the modes observed in a church bell or carillon bell is shown in Fig. 11.9. The relative modal frequencies and locations of the nodes are indicated. The columns represent modes having the same number of complete nodal meridians m. The end view at the bottom shows the configuration at the mouth of the bell, where $2m$ nodes are observed. The rows in the periodic table represent groups that have the same number n and configuration of nodal circles. The (2,0) or hum has no nodal circles and is assigned a group of its own. Group I includes the important family of modes having a nodal circle near the waist, while group II has a nodal circle near the mouth of the bell. The prime has a circular node about midway between the waist and the mouth, and thus, in a sense, it is a combination of the (2,1) mode of group I and the (2,1$^{\#}$) mode of group II.

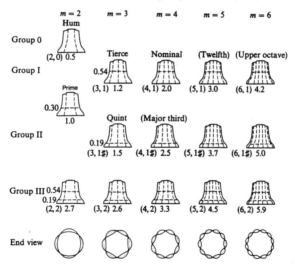

Fig. 11.9. Periodic table of inextensional modes of vibration in a church bell or carillon bell. Below each drawing are the modal frequencies of a D_5 church bell relative to the prime (which has essentially the same frequency as the strike note in a well-tuned bell). At lower left, (m,n) gives the number of nodal meridians and nodal circles n.

Vibrational frequencies of modes in groups 0-IX in a church bell with a D_5 strike note are shown in Fig. 11.10. Arrows denote the three partials in group I that determine the strike note (see Section 2.5). Graphical displays of several modes computed by the finite element method are shown in Fig. 11.11.

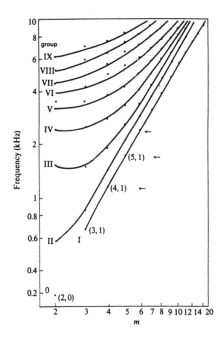

Fig. 11.10. Vibrational frequencies of groups O-IX in a D_5 church bell [8]. Arrows denote the three partials in group I that determine the strike note [8].

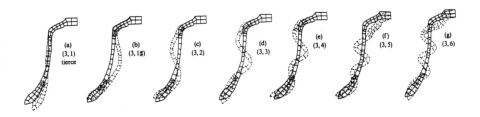

Fig. 11.11. Modal shapes in a church bell predicted by finite element calculation for the inextensional modes with $m=3$ [8].

In addition to the inextensional modes we have discussed, there are a number of vibrational modes that involve stretching of the bell. In these extensional modes, the radial and tangential motions are related by $v+\partial u/\partial \theta=0$ so that $u=\cos m\theta$ and $v=m \sin m\theta$. For $m=0$, in this case, the motion is entirely radial, but with increasing m, the tangential motion increases until it takes over from the radial motion. The modes for $m=0, 1$, and 2 are shown in Fig. 11.12. The $m=0$ mode can be described as a "breathing" mode.

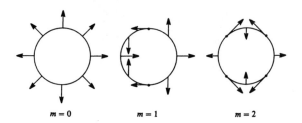

Fig. 11.12. Motion of a bell for extensional modes of small m.

11.3. Tuning and Temperament

Bell founders usually tune the lowest five modes of carillon bells so that their vibrational frequencies are in the ratios 1 : 2 : 2.4 : 3 : 4, and these ratios are used for tuned church bells as well. The bells are cast to a shape that gives nearly these ratios, and fine tuning is done by carefully thinning the inside of the bell at selected heights while it is mounted on a bell lathe. With the optimum shape, another five or six partials take on a nearly harmonic relationship, giving the bell a clear pitch and a musical quality. The various names of important partials are given in Table 11.1 along with the relative frequencies in an ideal bell (just tuning) and a bell with partials tuned to equal temperament.

Table 11.1. Names and relative frequencies of important partials of a tuned carillon bell or carillon bell [9]

			Ratio to prime (or strike note)		
Mode	Name of partials	Note name	Ideal (just)	Equal temperament	Bell in Fig. 11.1c
(2,0)	Hum	D_4	0.500	0.500	0.500
(2,1$^\sharp$)	Prime, fundamental	D_5	1.000	1.000	1.000
(3,1)	Tierce, minor third	F_5	1.200	1.189	1.183
(3,1$^\sharp$)	Quint, fifth	A_5	1.500	1.498	1.506
(4,1)	Nominal, octave	D_6	2.000	2.000	2.000
(4,1$^\sharp$)	Major third, deciem	$F_6^\sharp$	2.500	2.520	2.514
(2,2)	Fourth, undeciem	C_6	2.667	2.670	2.662
(5,1)	Twelfth, duodeciem	A_6	3.000	2.997	3.011
(6,1)	Upper octave, double octave	D_7	4.000	4.000	4.166
(7,1)	Upper fourth, undeciem	G_7	5.333	5.339	5.433
(8,1)	Upper sixth	B_7	6.667	6.727	6.796
(9,1)	Triple octave	D_8	8.000	8.000	8.215

Notice that the highest four partials in Table 11.1 are raised by as much as 4% above those of the ideal bell. A similar relationship is seen in other tuned church bells and carillon bells [10]. This stretching of the partial series may very well contribute a desirable quality to the bell sound [11].

Not all church bells have harmonically tuned partials, however. A study of 363 church bells in Western Europe revealed that only 17% have the hum, the prime, and the nominal tuned in octaves [12]. The distribution of the various tunings is given in Table 11.2. Terhardt and Seewan [13] also found a rather wide distribution in the tuning of partials in historic German bells.

Table 11.2. Distribution of various tunings among Western European church bells [12]

	Tonal Structure			
Bell type	Hum	Fundamental	Nominal	%
Octave bell with perfect fundamental	C_4	C_5	C_6	17.4
Octave bell with diminished fundamental	C_4	B_4	C_6	14.3
Minor-ninth bell with perfect fundamental	B_3	C_5	C_6	7.4
Minor-ninth bell with augmented fundamental	B_3	$C^\sharp_5$	C_6	7.2
Octave bell with augmented fundamental	C_4	$C^\sharp_5$	C_6	6.9
Major-seventh bell with perfect fundamental	$C^\sharp_4$	C_5	C_6	5.8
Major-seventh bell with diminished fundamental	$C^\sharp_4$	B_5	C_6	5.5
Minor-ninth bell with diminished fundamental	B_3	B_4	C_6	4.7
Major-seventh with double diminished fundamental	$C^\sharp_4$	$B^\flat_4$	C^6	4.1
Octave bell with double diminished fundamental	C_4	$B^\flat_4$	C_6	3.6
Minor-seventh bell with diminished fundamental	D_4	B_4	C_6	2.7
Minor-seventh bell with double diminished fundamental	D_4	$B^\flat_4$	C_6	2.5
Minor-seventh bell with triple diminished fundamental	D_4	A_4	C_6	2.5

11.4. The Strike Note

When a large church bell or carillon bell is struck by its metal clapper, one first hears the sharp sound of metal on metal. This metallic strike sound includes many inharmonic partials that die out quickly, giving way to a strike note or strike tone that is dominated by the prominent partials of the bell. Most observers identify the strike note in a tuned bell as having a pitch at or near the frequency of the strong second partial (prime or fundamental), but to others its pitch is an octave higher. Finally, as the sound of the bell ebbs, the slowly decaying hum tone (an octave below the prime) lingers on. An historical account of research on the strike note is given in ref. 14.

The strike note is of great interest to psychoacousticians, because it is a subjective tone created by three strong harmonic (or nearly harmonic) partials in the bell sound. The octave or nominal, the twelfth, and the upper octave normally have frequencies nearly in the ratios 2:3:4 (see Table 11.1). The ear assumes these to be partials of a missing fundamental, which it hears as the strike note (see Section 2.5).

Scientists at the Institute of Perception Research in The Netherlands did an experiment to ascertain the role of each partial in determining the pitch of the strike note. The sound of a Hemony bell was recorded, and by means of a digital filter, each of the first nine partials was raised and lowered in frequency up to 10% while listeners judged the pitch of the resulting strike note in a pitch-matching experiment. The results (Fig. 11.12) show that partials five and six (the octave and the twelfth) are the most important, followed by partial seven (the upper octave). The other partials, including partial two (the prime, which coincides closely to the strike note in frequency), have very little effect on the pitch of the strike note, as indicated on the vertical axis [15].

In very large bells, a secondary strike note may occur a musical fourth above the primary strike note and may even appear louder under some conditions [16]. This secondary strike note is a subjective tone created by four partials beginning with the upper octave. These partials correspond to the (6,1), (7,1), (8,1), and (9,1) modes of vibration, whose frequencies are nearly three, four, five, and six times that of the secondary strike note (see Table 11.1). In a large bell (800 kg or more), these partials lie below 3000 Hz where the virtual pitch is quite strong [17].

In small bells, the higher partials lie at a very high frequency where the virtual pitch is weak, and so both the primary and secondary strike notes are weak. The pitch is then determined mainly by the hum, the prime, and the nominal, which normally are tuned in octaves. It is sometimes difficult to decide in which octave the pitch of a small bell lies, especially if the frequencies of these three partials are not exactly in 1:2:4 ratio.

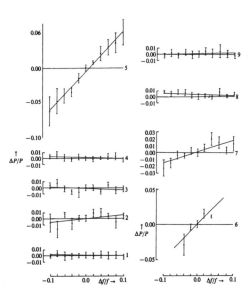

Fig. 11.12. Effect of partials 1 to 9 in determining the strike note of a bell. $\Delta P/P$ is the relative change in pitch resulting from a change in the relative frequency $\Delta f/f$ of a partial. Note the great importance of the fifth and sixth partials (octave and twelfth) followed by the seventh partial (upper octave); other partials are relatively unimportant [15].

11.5. Major-Third Bells

For years, some carillonneurs have felt that a composition in a major key, especially if it includes chords with many notes, might sound better if played on bells with a major character, which suggests replacing the strong minor-third partial with a major-third partial. Efforts to fabricate such bells were not successful, however, because raising the minor-third partial by changing the profile also changed the other members of the $(m,1)$ family.

Employing a technique for structural optimization using finite element methods on a digital computer, scientists at the Technical University in Eindhoven and the Eijsbouts Bellfoundry in The Netherlands were able to design bells with the minor-third partial replaced by a major-third. The first bell, introduced in 1985, had rather rapidly decaying partials, but a second bell had a longer decay, more nearly that of the traditional minor-third bell [12]. Both major-third bells had a rather large bulge near the waist of the bell, and were rather easily identified visually, as can be seen in Fig. 11.13. In 1999, a new major-third bell design, which avoids the bulge, was announced.

Fig. 11.13. Two major-third bells (left) compared to conventional minor-third bells with the same pitches [18].

11.6. Scaling of Bells

In Chapter 8, it was shown that bending waves in a thin plate travel at a speed that is proportional to the plate thickness. In a circular cylinder with free ends, the inextensional (bending) modes will have frequencies that are proportional to the thickness and inversely proportional to the square of the radius. To a first approximation, then, it should be possible to scale a set of bells of the same material and with similar profiles according to the scaling law $f=Ch/D^2$, where h is the thickness, D is the diameter, and C is a constant that depends on the material and the profile.

Insofar as the simple scaling law for cylinders applies to bells, it should be possible to scale a set of bells by making all dimensions proportional to $1/f$. This scaling approximates that found in many carillons dating from the fifteenth and sixteenth centuries [4]. However, since a $1/f$ scaling causes the smaller treble bells to have a rather weak sound, later bell founders increased the sizes of their treble bells [19]. The average product of frequency times diameter in several fine seventeenth century Hemony carillons has been found to increase from 100 m/s (in bells of 30 kg or larger) to more than 150 m/s in small treble bells. The measurements on the bells in three Hemony carillons are combined in Fig. 11.13. The straight, solid lines indicate a $1/f$ scale. Deviations from this scaling in the treble bells are apparent.

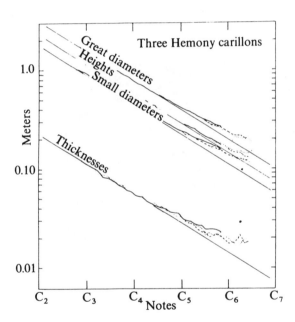

Fig. 11.14. Measurements of the bells in three Hemony carillons: Oude Kerk, Amsterdam, François Hemony, 1658; Dom Toren, Utrecht, François and Pieter Hemony, 1663; and Heilige Sulpiciuskerk, Diest, Pieter Hemony, 1670. The solid lines represent a $1/f$ scaling law [19].

Studies of bells cast at the Eijsbouts bell foundry indicate that a $1/f$ scaling is used for swinging bells, but for carillon bells the diameter of the high-frequency bells is substantially greater than predicted by this scaling [20]. This is probably the most notable difference between carillon bells and tuned church bells.

Actually, the simple scaling law for thin cylinders needs modification for bells which have relatively thick walls. In the modes of group I, which have antinodes near the soundbow, the frequency is roughly proportion to $h^{0.7}$, where h is the thickness, while in modes of group II, which have antinodes near the waist, the frequency is roughly proportional to $h^{0.86}$, which is nearer the $f \propto h$ behavior in flat plates and circular cylinders [10]. Fig. 11.15 shows how the frequencies of several partials in a bell vary with thickness, according to calculations using finite element methods [21]. A more refined version of the scaling law which takes this into account is $f=Ch^{\beta}/D^{1+\beta}$, where $\beta=0.615$ for the hum note and takes on other values (but always less than one) for other partials [21].

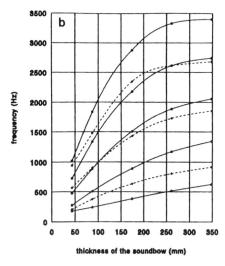

Fig. 11.15. How partial frequencies of a bell vary with thickness of the soundbow. Solid lines are for partials without a nodal circle (Group I), while dashed lines are for partials with a nodal circle (Group II) [21].

For the bell founder, it is important to know the mass of each bell. That can be determined from the formula: $M=K(fD)^{-\beta}D^3$ where f is its frequency, D its diameter, and K is the mass constant for the particular profile chosen.

The effect of wall thickness on the partials of a bell is illustrated in Table 11.3 by the partial frequencies of three experimental bells of the same diameter with thickness 2/3 normal, normal, and 3/2 normal [22]. The C_5 bell of normal thickness (with a profile constant $K=0.323$ kg•s/m^4) has the partials whose note names are given in the center column. To the left are shown the partials of the thin bell and to the right are the partials of the thick bell. The raising and lowering are shown on a scale of cents. Note that the harmonic relationships between partials have been altered considerably by changing the thickness. This is also the case with bells whose thickness has been altered only slightly [10].

Table 11.3. Effect of wall thickness on the partials of a C_5 bell (adapted from ref. 22)

Change in frequency (cents)	1000	900	800	700	600	500	400	300	200	100	0	100	200	300	400	500	600
Hum		E_4									C_5						F_5
Prime					$F^\sharp_5$						C_6		$D^\sharp_6$				
Minor third			G_5								$E^\flat_6$					$G^\sharp_6$	
Fifth	A_5										G_6						$C^\sharp_7$
Octave			$D^\sharp_6$								C_7					F_7	
Major third			$F^\sharp_6$								E_7						$A^\sharp_7$
Fourth							$C^\sharp_7$				F_7-		G_7				
Fourth				A_{6+}							F_7					$A^\sharp_7-$	
Upper fifth			$A^\sharp_{6+}$								G_7						C_8
Thickness				2/3							1/1					3/2	

11.7. Sound Decay and Warble

A vibrating bell loses energy mainly by sound radiation, although internal losses also play a role. The sound pressure level of each radiated partial decays at a constant rate as the vibrational energy decays exponentially, and thus it is customary to express the 60-dB decay time for each main partial.

The decay times for the principal modes of vibration of the D_5 church bell described in Fig. 11.9 are given in Table 11.4. Note the long decay time of the (2,0) mode (hum) and the relatively short decay times of the modes of higher frequency. This is mainly due to the greater radiation efficiency of the higher modes. Schad and Warlimont [23] found that the damping due to internal losses was approximately the same for all the principal modes, so the large differences in decay times are indicative of different rates of radiation.

Some bells exhibit a pronounced amplitude modulation or "warble," which is caused by the beating together of the nearly degenerate components of a mode doublet. Warble can occur in any partial, and the warble frequency is equal to the frequency difference between the doublet pair. In theory, warble can be eliminated by selecting the strike point to lie at a node for one component and an antinode for the other. In practice, this may not be useful because doublet splitting is likely to occur in several doublet pairs and selecting the strike point to minimize warble in one pair may enhance it in another. Various methods of trying to guarantee the correct alignment of doublet pairs by suitable breaking of the axial symmetry have been suggested, including, for example, the addition of two diametrically opposite meridian ribs at whose location the clapper strikes [24].

11.8. Sound Radiation

The most prominent partials in the spectrum of a church bell or carillon bell are radiated by the $(m,1)$ modes belonging to group I. These modes can be considered to be due to standing flexural waves. With the exception of the $(2,0)$ mode (hum), all modes have a nodal circle about halfway up the bell. For the purpose of understanding the general properties of the radiation field of the bell, we can roughly model its outside surface as a collection of $4m$ sources alternating in phase ($2m=4$ sources in the case of the $(2,0)$ mode). Another $4m$ sources of slightly smaller size occur on the inside surface.

The radiation efficiency of such a collection of alternating sources increases with frequency and with the size of the bell. Increasing the diameter of a bell (and at the same time increasing the thickness to keep the frequency the same) increases the area of each source and thus increases the radiation efficiency. However, a more significant increase in radiation efficiency occurs when the separation between adjacent sources of opposite phase exceeds a half-wavelength of sound in air. Another way to express this condition is that when the speed of flexural waves in the bell exceeds the speed of sound in air, radiation efficiency increases markedly; this occurs at the so-called *coincidence frequency* (sometimes called the critical frequency). The speed of flexural waves in a plate is given by $v(f) = \sqrt{1.8 c_L h f}$, where h is the thickness, f is the frequency, $c_L = \sqrt{E/\rho(1-v^2)}$ is the longitudinal wave speed, E is Young's modulus, v is Poisson's ratio, and ρ is the density.

The flexural wave speed in a bell is roughly the mode frequency times the circumference divided by mode number m. For the D_5 church bell in Fig. 11.9, having a diameter of 70 cm, the flexural wave speed is roughly $(292.7)(0.7\pi/2) = 322$ m/s for the $(2,0)$ mode, but it increases to about 644 m/s for the $(4,1)$ mode and to about 1180 m/s for the $(9,1)$ mode. Since all but the lowest mode exceed the speed of sound in air (343 m/s at 22° C), the bell radiates most of its partials quite efficiently.

11.9. Clappers

The sound of a bell is very much dependent on the size, shape, and hardness of the clapper, the point at which it strikes the bell, and the strength of the blow. Not much systematic research on clappers has been reported in the literature, however.

Church bell clappers were formerly made of wrought iron, which probably remains the preferred material, although in recent years it has been gradually replaced by cast iron. In change ringing, which is especially popular in England, bells are rung full circle (that is, the bell is turned completely over). Clappers are flung against the bells with considerable force, and they must be carefully designed to avoid breaking. Clappers in carillon bells are usually made of steel.

Bigelow [19] reports the results of an experiment in which an F_5 carillon bell was struck by three different clappers of varying weights, each falling from three different heights. A heavy hammer was found to increase the strength of the lower partials, but to decrease those of the nominal (or octave). Since the nominal contributes in an essential way to the subjective strike note, it is implied that a heavy hammer will tend to diminish the intensity of that note. It should be noted that in a larger bell the nominal is usually found to be the most prominent partial for the first few seconds after the strike.

11.10. Bell Metal

Bronze (nominally 80% copper and 20% tin) has been the preferred metal for bells since ancient times, because of its hardness and low damping. Technology for the casting of bronze articles was well established as long ago as 2000 BC. Bronze could be melted in a charcoal fire, and its pouring temperature of only 1100°C did not impose severe requirements on the mold material. There is, indeed, a close parallel between the manufacture of bells and cannon, and very similar materials and processes were used throughout history.

Schad and Warlimont [23] reported an extensive study of the effects of variations in bronze composition damping and other properties of bronze. Increase in percentage of tin greatly reduces the internal damping but increases the sound velocity, while addition of lead up to 10% reduces the hardness, increases the damping, and marginally reduces the sound velocity. One concludes that the composition of standard bell bronze is close to optimal, but that close attention must be paid during casting to minimizing porosity, which can markedly increase internal damping.

Church bells have been cast of steel and aluminum, which are less expensive than bronze, but the resulting sound is generally judged inferior to bell bronze.

References
1. A. Lehr, *The Art of the Carillon in the Low Countries* (Lannoo, Tielt, Belgium, 1991).
2. Lord Rayleigh, *The Theory of Sound*, vol I (Macmillan, New York, 1894). (Reprinted by Dover, New York 1945).
3. A. B. Simpson, *Nash's Pall Mall Magazine* **7**, 183-194 (1895).
4. E. W. van Heuven, *Acoustical Measurements on Church Bells and Carillons (De Gebroeders van Cleef, s'Gravenhage, 1949).*
5. J. R. Lawson, *Bulletin, Guild of Carillonneurs in North America* **35**, 19-22 (1986).
6. F. G. Tyzzer, *J. Franklin Inst.* **210**, 55-56 (1930).
7. T. D. Rossing, *American Scientist* **72**, 440-447 (1984).
8. R. Perrin, T. Charnley, and J. DePont, *J. Sound Vibr.* **90**, 29-49 (1983).
9. T. D. Rossing and R. Perrin, *Applied Acoustics* **76**, 1263-1267 (1987).
10. A. Lehr, *J. Acoust. Soc. Am.* **79**, 2000-2011 (1986).
11. F. H. Slaymaker, *J. Acoust. Soc. Am.* **47**, 1569-1571 (1970).
12. A. Lehr, *The Design of Swinging Bells and Carillon Bells in the Past and Present* (Athanasius Kircher Foundation, Asten, The Netherlands, 1987).
13. E. Terhardt and M. Seewann, *Acustica* **54**, 129-144 (1984).
14. T. D. Rossing, *Acoustics of Bells.* (Van Nostrand-Reinhold, Stroudsburg, PA, 1984).
15. J. H. Eggen, *The Strike Note of Bells* (Institute for Perception Research, Eindhoven, The Netherlands, 1986).
16. J. F. Schouten and J. 'T Hart, *Die slagtoon van klokken* (Netherlands Acoustical Society, 1965). (English translation in ref. 14).
17. R. J. Ritsma, *J. Acoust. Soc. Am.* **42**, 191-198 (1967).
18. N. H. Fletcher and T. D. Rossing, *The Physics of Musical Instruments*, 2nd ed. (Springer-Verlag, New York, 1998).

19. A. L. Bigelow, *The Acoustically Balanced Carillon* (School of Engineering, Princeton, New Jersey, 1961).
20. A. Lehr, *Acustica* **2**, 35-38 (1952).
21. A. Lehr, *Acustica* **83**, 320-336 (1997).
22. A. Lehr, *Leerboek der Campanologie* (National Beiaardmuseum, Asten, The Netherlands, 1976).
23. C.-R. Schad and H. Warlimont, *Acustica* **29**, 1-14 (1973). (English translation in [14]).
24. R. Perrin, T. Charnley, and H. Banu, *J. Sound Vibr.* **80**, 298-303 (1982).

Chapter 12
Handbells, Choirchimes, Crotals, and Cowbells

Handbells date back to at least several millennia BC. Chinese legends mention them in the 27th century BC, and the recorded history of bells in China goes back to the Shang dynasty (1600 BC). Bells existed in the Near East and India before 1000 BC, and they were common in the ancient civilizations of Egypt, Greece, and Rome.

Bells, large and small, were adopted by Christianity as early as 400 AD. Nola, a city in the Campania region of Italy and a center for bronze casting, furnished small bells for use in Christian worship. In the National Museum of Ireland in Dublin is a bell said to have been carried by St. Patrick in the 5th century and used to perform miracles. A bronze handbell from 963 AD in the Museo Provinciale in Cordoba, Spain, is said to have been modeled after the helmet of a Roman soldier.

Bells developed as Western musical instruments in the seventeenth century when bell founders discovered how to tune their partials harmonically. Tuned handbells developed in England in the eighteenth century. One early use of handbells was to provide tower bellringers with a convenient means to practice change ringing. In more recent years, handbells choirs have become popular in schools and churches; some 40,000 choirs are reported in the United States alone. A set of handbells used by a handbell choir is shown in Fig. 12.1.

Fig. 12.1. A seven-octave set of handbells used by a modern handbell choir.

Handbells

12.1. Vibrational Modes of Handbells

A periodic table of vibrational modes in a handbell are shown in Fig. 12.2. The modes are somewhat similar to those of a church bell or carillon bell shown in Fig. 11.9, although certain differences can be noted. Group 0, consisting of modes with no nodal circles, includes both a (2,0) and a (3,0) mode; in larger handbells a (4,0) mode and even a (5,0) mode are present as well. When Group 0 includes a (4,0) mode, group I begins with the (5,1) mode and group II with the $(5,1^\#)$ mode.

Relative modal frequencies in Fig. 12.2 are for a medium-size C_5 handbell; other handbells will have different modal frequencies, except for the (3,0) mode which is nearly always tuned to 3 times the fundamental (2,0) frequency.

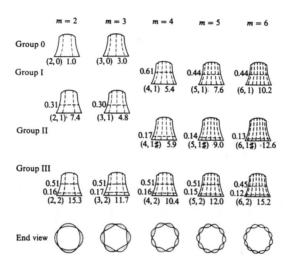

Fig. 12.2. Periodic table of vibrational modes in a handbell. Below each drawing are the relative modal frequencies in a Malmark C_5 handbell. At lower left, (m,n) gives the number of nodal meridians $2m$ and nodal circles n [1].

The vibrational frequencies of a C_5 handbell are shown in Fig. 12.3. A comparison with Fig. 11.10 for a church bell or carillon bell shows some interesting similarities and differences. Note that each curve in Fig. 12.3 drawn for $n=1, 2, 3,...$ shows a minimum in frequency at about $m=n+2$. This is similar to the behavior of a cylinder with a fixed end cap, in which the stretching energy becomes substantial for small m [2]. Thus, the total strain due to stretching and bending in a cylindrical shell with fixed ends decreases with m until it reaches a minimum, then increases with m [3]. In large handbells (G_3 and larger), the minimum occurs at about $m=n+3$.

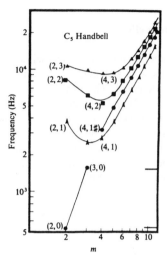

Fig. 12.3. Vibrational frequencies of a C_5 handbell [4].

Hologram interferograms of a number of modes in the C_5 handbell are shown in Fig. 12.4. The bull's eyes locate the antinodes. Note that the upper half of the bell moves very little in the (7,1) mode; the same is true in $(m,1)$ modes when $m>7$. The modes of vibration calculated in a C_5 handbell using finite element methods were found to be in good agreement with the modes observed by hologram interferometery [4].

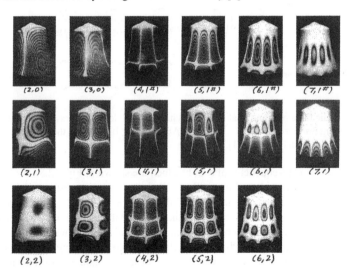

Fig. 12.4. Hologram interferograms of mode shapes in a C_5 handbell [4].

12.2. Sound Radiation

The radiation efficiency of a bell, as we discussed in Section 11.8, depends on the relative speed of flexural waves compared to the speed of sound in air. This generally occurs above the coincidence frequency, which is determined by the thickness and the elastic properties of the metal. A thick church bell radiates most of its partials quite efficiently, but a thin handbell does not, since the speed of the flexural waves is considerably less than the speed of sound in air. Air adjacent to the vibrating surface tends to flow back and forth between adjacent areas of opposite phase, creating a sort of acoustic short circuit.

In the C_5 handbell, for example, the flexural wave speed is roughly $(523)(0.06\pi) = 100$ m/s for the fundamental (2,0) mode. Since this is considerably less than the speed of sound (about 343 m/s at room temperature), sound radiation is not very efficient. The (3,0) and $(4,1^\#)$ modes in the same handbell have flexural wave speeds of approximately 200 m/s and 300 m/s, respectively, and thus these modes tend to radiate sound a little more efficiently than the fundamental (2,0) mode. In large handbells, the radiation efficiency is low, because the flexural wave velocity for all the principal modes is less than the speed of sound in air. In a G_2 handbell, for example, the flexural wave speeds are only 41 m/s and 71 m/s for the (2,0) and (3,0) modes, respectively [6]. This is a significant problem in handbell design.

In addition to the direct radiation of sound normal to its vibrating surfaces, a bell also radiates sound axially at twice the frequency of each vibrational mode. The mechanism for this radiation is shown in Fig. 12.5. In the (2,0) mode, as the mouth of the bell changes from a circle to an ellipse, some air is expelled from the mouth due to a net change in volume. The radiated sound thus includes a second harmonic. Similarly, the (3,0) mode, which is tuned to the 3rd harmonic, radiates a 6th harmonic. The intensity of this axially radiated sound increases with the fourth power of the vibrational amplitude, whereas the direct radiation increases only with the square of the amplitude [5].

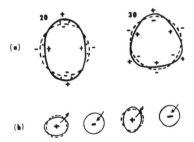

Fig. 12.5. Sound radiation from a vibrating bell. (a) Direct radiation from the (2,0) and (3,0) modes; regions of positive and negative pressure change are shown. (b) Radiation at twice the frequency of the (2,0) mode due to net volume change; a similar radiation occurs at twice the frequency of the (3,0) mode [5].

The fundamental (2,0) mode in a handbell radiates a fairly strong second harmonic partial along the axis, as well as a fundamental whose maximum intensity is perpendicular to the axis. The (3,0) mode also radiates at twice the vibrational frequency, but the sixth harmonic partial is quite weak except in very large handbells. Thus, in most handbells, the principal harmonic partials are the first, second, and third harmonics.

12.3. Sound Decay and Warble in Handbells

Sound decay curves for a large and a small handbell are shown in Fig. 12.6. Note that the small handbell decays much faster than the large one. This is partly due to the more efficient radiation of sound by the smaller bell (which is operating closer to its coincidence frequency). Note that the second harmonic decays at twice the rate of the fundamental, because its intensity is proportional to the square of the amplitude of the (2,0) mode, while the fundamental intensity is directly proportional to this amplitude.

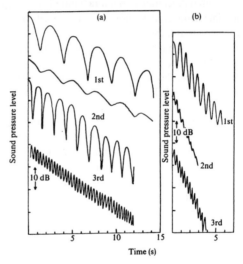

Fig. 12.6. Sound decay from: (a) G_4 handbell; (b) C_6 handbell. The first and second harmonics are radiated by the (2,0) mode, the third harmonic by the (3,0) mode.

A prominent feature in Fig. 12.6 is the warble that results from the beats between two members of the doublet mode which have slightly different frequency. This warble is greatest at angles that lie midway between the maxima of the two mode members. In "voicing" a handbell, the clapper is set to strike the bell at one of these two maxima so as to minimize warble. Unfortunately, the optimum strike point for the upper partials may not coincide with that of the fundamental mode.

12.4. Timbre and Tuning of Handbells

The sound of a handbell is very different from that of a carillon bell or church bell, even though they are cast from the same bronze alloy. Church bells and carillon bells are struck by heavy metal clappers in order to radiate a sound that can be heard at a great distance, whereas the gentle sound of a handbell requires a relatively soft clapper. Handbell choirs play chords of many notes, rapid passages, and staccato notes, all of which favor tones that are dominated by the fundamental and not too rich in higher partials.

In the so-called English tuning of handbells, followed by most handbell makers in England and the United States, the (3,0) mode is tuned to three times the frequency of the (2,0) mode. The fundamental (2,0) mode radiates a rather strong second harmonic partial, however, so that the sound spectrum has prominent partials at the first three harmonics [5]. Some Dutch founders aim at tuning the (3,0) mode in handbells to 2.4 times the frequency of the fundamental, giving their handbell sound a minor-third character somewhat like a church bell. Such bells are usually thicker and heavier than bells with English-type tuning [6].

A handbell, unlike a church bell, appears to sound its fundamental pitch almost from the very onset of sound. There are several reasons for the absence of a separate strike note. First of all, there is no group of harmonic partials to create a strong subjective tone. Secondly, handbells employ a relatively soft, nonmetallic clapper, so that there is no sound of metal on metal, and the partials develop a little more slowly after the clapper strikes the bell.

Handbell clappers strike the bell with a soft surface of plastic, leather, or felt. In most handbell clappers, surfaces of different hardness can be selected in order to vary the timbre of the sound. Figure 12.7 illustrates how different clapper surfaces change the balance between partials and hence the timbre. In general, the softer clapper favors the fundamental. Note the difference between sound radiated along the bell axis and that radiated at right angles to the bell.

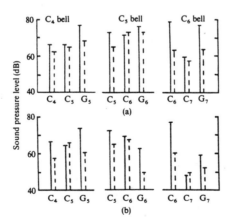

Fig. 12.7. Comparison of the principal partials of handbell sound radiated along the bell axis (a) and perpendicular to the axis (b) with hard (—) and soft (- - -) clappers, respectively. Sound pressure levels were recorded 1 m from each handbell in an anechoic room [5].

It is interesting to note how much the relative modal frequencies vary from large to small bells, as shown in Fig. 12.8. The $(m,1^\#)$ families change relatively little from small to large bells, but the $(m,1)$ families show considerable change. The (2,1) mode, in particular varies from 5 times the fundamental in a small C_6 bell to 23 times the fundamental in the G_1. In a large church bell, by contrast the prime (which is radiated by the mode most closely related to the $(2,1^\#)$ handbell mode, has twice the frequency of the hum (2,0).

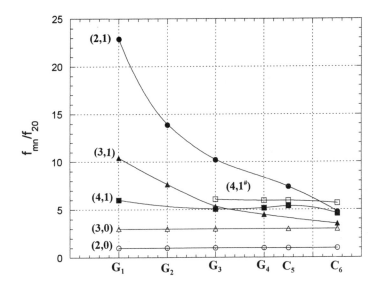

Fig. 12.8. Relative frequencies of (2,0), (3,0), $(4,1^\#)$, (2,1), (3,1) and (4,1) modes in handbells.

12.5. Scaling of Handbells

Handbells are generally cast much thicker than the final bell, so that a considerable quantity of metal is removed on the tuning lathe. In fact, the same size of casting can be used for more than one bell by adjusting the wall thickness of the final bell. Under these conditions, the scaling curves for sets of handbells might be expected to be less regular than the scaling curves for church bells, such as the one shown in Fig. 11.14.

The general scale followed by most handbell makers is to make the diameter inversely proportional to the square root of frequency, except for the smallest bells in which it varies inversely with the cube root of frequency instead [5]. The thickness is then adjusted so that h/d^2 is nearly proportional to the frequency. The scaling of a five-octave set of Malmark handbells is shown in Fig. 12.9.

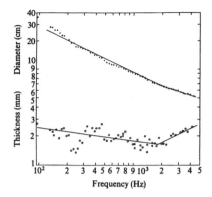

Fig. 12.9. Scaling of a five-octave set of Malmark handbells [5].

12.6. Bass Handbells

Demand for handbells of lower and lower pitch has led to development of bass bells as low as G_o (fundamental frequency of 24.5 Hz). Unfortunately, these large bass bells radiate inefficiently, especially bells made of bronze, because the speed of bending waves is well below the speed of sound, and therefore they operate well below the conicidence frequency.

In order to obtain a higher radiation efficiency and thereby enhance the sound of bass bells, the Malmark company makes a bass handbell of aluminum. These aluminum bells are larger in diameter than the corresponding bronze bells, and they have lower coincidence frequencies, both of which lead to more efficient radiation of bass notes. In addition, they are considerably lighter in weight, and thus they are much more easily handled by bell ringers [8]. A G_1 aluminum handbell is shown in Fig. 12.10.

Fig. 12.10. G_1 aluminum bass handbell.

Holographic interferograms of vibrational modes in an aluminum handbell are shown in Fig. 12.11. Note that the $(m,0)$ family ends with the $(5,0)$ mode and thus the $(m,1^\#)$ family begins with the $(6,1^\#)$ mode. Although not shown in Fig. 12.11, modes up to $(15,5)$ were observed in these studies.

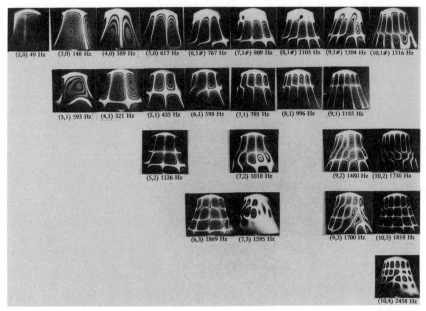

Fig. 12.11. Holographic interferograms of vibrational modes in a G_1 aluminum bass handbell [8].

Sound spectra from large bass handbells (both of aluminum and bronze) are characterized by a strong sixth harmonic partial not readily observed in smaller handbells. This partial is radiated by the (3,0) mode in the manner described in section 12.2, just as the second harmonic is radiated by the (2,0) mode in nearly all handbells.

12.7. Choirchimes

The Choirchime, developed by Malmark, Inc., a manufacturer of handbells, is essentially a closed-end self-resonant tuning fork with a handbell clapper. Now available in chromatic sets up to 5 octaves, Choirchimes have become very popular in schools and churches, both for teaching and performing music. Choirchimes are presently available from G_2 (98 Hz) to C_8 (4186 Hz). The closed-tube design used in Choirchimes keeps the lower-pitched chimes to a manageable length.

12.7.1. Modes of Bending Vibration

The vibrational modes of an ordinary tuning fork include: (1) symmetrical bending modes in the plane of the fork; (2) antisymmetrical modes in the plane; (3) symmetrical bending modes out of the plane [9]. In the symmetrical modes, each tine moves in the manner of a cantilevered beam, while in the antisymmetrical modes the fork moves in the manner of a beam with free ends. Thus, the modal frequencies of the symmetrical mode would vary as $(2n+1)^2$, with $n = 1, 2, \ldots$. For the in-plane modes, the modal frequencies of a tuning fork fit this simple model quite well [9].

In Choirchimes, we would also expect to find symmetrical and antisymmetrical bending modes, both in the plane parallel to the slotted faces and perpendicular to this plane. As in the case of an ordinary tuning fork, the symmetrical modes might be modeled by two cantilevered beams, whose modal frequencies are given by :

$$f_n = \frac{\pi K}{8L^2}\sqrt{\frac{E}{\rho}}[(1.194)^2, (2.988)^2, 5^2, 7^2, \ldots (2n-1)^2], \qquad (12.1)$$

where E is Young's modulus, ρ is the density, L is the length of the tines, and K is the radius of gyration of the beam cross section. The antisymmetrical modes, on the other hand, might be modeled by a beam with free ends, whose modal frequencies are given by:

$$f_n = \frac{\pi K}{8L^2}\sqrt{\frac{E}{\rho}}[(3.011)^2, 5^2, 7^2, \ldots (2n+1)^2]. \qquad (12.2)$$

Of course these models only approximate the real behavior. In the case of the symmetrical modes, the attached ends of the "tines" are not completely clamped, and in the case of the antisymmetric modes, the slotted end has less mass and less bending stiffness than the other end.

Bending mode frequencies for an A_3 and an A_4 Choirchime are shown in Fig. 12.12. The symmetric modes are plotted vs $2n$-1, as in equation 12.1, and the parallel modes are plotted vs $2n$+1, as in equation 12.2. All four lines have slopes of 2. Both types of modes show essentially a quadratic dependence on the appropriate factor $2n$-1 or $2n$+1, as predicted by the theory of bending waves in thin beams.

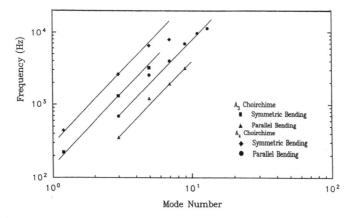

Fig. 12.12. Frequencies of bending modes in two Choirchimes. Symmetrical modes are plotted vs $2n$-1, antisymmetic modes vs $2n$+1. Solid lines have a slope of 2 [10].

Bending mode frequencies in 5 Choirchimes are given in Table 12.1. Ratios to the fundamental frequency f_1 are also given. The higher modes are not harmonically related to the fundamental, but that is relatively unimportant since the sound is dominated by the fundamental.

Table 12.1. Bending mode frequencies in Choirchimes and ratios to the fundamental f_1

	f_1	f_2	f_3	f_4	f_2/f_1	f_3/f_1	f_4/f_1
A_3	220	349	1208	1310	1.59	5.49	5.95
A_4	440	688	2559	2619	1.56	5.82	5.95
E_5	659	976	2803	3935	1.48	4.25	5.97
C_6	1046	1359	3861	6063	1.30	3.69	5.80
G_6	1569	2057	5815	8772	1.31	3.71	5.59

The parallel bending modes in the C_6 and G_6 Choirchimes are proportionally lower in frequency and are scaled longer for the convenience of the player. Since Choirchimes employ closed-tube resonators, the tubes can extend as far beyond the closed end as desired without affecting the resonator tuning. From a musical point of view, the tuning of these higher modes is relatively unimportant. They are not tuned to harmonics of the fundamental, and the parallel bending modes decay rapidly when the chimes are held in the player's hand.

12.7.2. Modes of Torsional Vibration

Two types of torsional modes would be expected: those in which the tines twist in opposite directions, so that the uncut portion of the tube remains stationary; and those resembling the torsional modes of a beam with free ends in which the tines twist in the same direction. Torsional mode frequencies of a beam with a square cross section having free ends are given by:

$$f_n = 0.92 \frac{n}{2L} \sqrt{\frac{G}{\rho}}, \quad n = 1, 2, 3, \ldots, \qquad (12.3)$$

where the shear modulus G is related to Young's modulus E and Poisson's ratio ν by the equation $G = E/2(1+\nu)$. If one end of the bar is clamped (as is approximately true when the tines bend in opposite directions), the mode frequencies are given by:

$$f_n = 0.92 \frac{n}{4L_t} \sqrt{\frac{G}{\rho}}, \quad n = 1, 3, 5, \ldots, \qquad (12.4)$$

where L_t is the effective tine length.

Holographic interferograms of three parallel torsional modes and three symmetric torsional modes are shown in Fig. 12.13. These modes radiate very little sound.

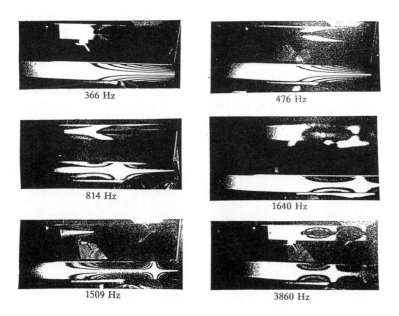

Fig. 12.13. Torsional modes in an A_3 Choirchime. In the parallel modes at the left, the tines rotate in the same direction; in the symmetrical modes at the right, the tines rotate in opposite directions [10].

12.7.3. Sound Spectra

Sound spectra are dominated by the fundamental, giving the choirchime its characteristically "pure" tone. Partials due to the first parallel bending mode (f_2 in Table 12.1) and the second symmetrical bending mode (f_3) are at least 20 dB (A_4 chime) to 30 dB (A_3 chime) lower in sound level at the time of striking. After 0.1 s, these partials are more than 40dB lower than the slowly-decaying fundamental [10].

The difference in sound level with and without the resonator ranges from 20 to 30 dB with Choirchimes of different size. This dramatic increase in sound level is unusual for musical instrument resonators. The tubular resonators of a marimba or xylophone, by way of comparison, add about 3 to 6 dB to the sound level (see Chapter 6). The dramatic increase in sound level is due to the fact that the resonator acts essentially as a monopole source and therefore radiates very efficiently, whereas the tines themselves act more nearly as a quandrupole source and radiate with much lower efficiency (see section 2.7). It is probably fair to say that a Choirchime produces the "purest" tone (i.e., free of overtones) of any musical instrument known.

12.8. Chinese Qing

The *qing* (also known as the *shun* or *ching*) is a bowl-shaped musical instrument, commonly used in Buddhist religious ceremonies, where it is often paired with a *mu yu* or wooden fish of about the same size (as shown in Fig. 12.14). *Qing* generally range from 10 to 40 cm in diameter and 8 to 35 cm in height, although one large *qing* from the Han dynasty (206 BC - 210 AD) measures 75 cm in diameter [11]. When used in religious ceremonies, the *qing* generally rests on a silk pillow and is struck at the rim with a wooden stick. Figure 12.15 shows four bronze *qing* from 10 to 18 cm in diameter that we have studied in our laboratory. In ancient times, the *qing* was often engraved with the text of the Buddhist Sutra, whose wonders would be conveyed by the sound of the instrument.

Fig. 12.14. Scene from a Buddhist temple, showing a large *qing* (right) and a *mu yu* (left).

Fig. 12.15. Four *qing* with diameters of 18, 15, 12, and 10 cm.

The prinicpal modes of vibration result from the propagation of bending waves around the circumference. Viewed in the axial direction, these modes resemble those of a bell, with the $(m,0)$ mode having $2m$ nodes around the mouth. The $(m,1)$ and higher families of bell modes are not observed, however. Figure 12.16 shows holographic interferograms of some of the more prominent vibrational modes of an 18-cm diameter *qing* (the largest one in Fig. 12.15). Modes (2,0) through (9,0) are identifiable in the top two rows, but it is difficult to assign mode numbers at the high frequencies.

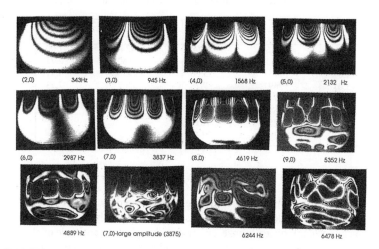

Fig. 12.16. Holographic interferograms of an 18-cm diameter bronze *qing*, showing modal shapes of the $(m,0)$ modes (top two rows). Modes in the bottom row are not identified, except for the second one, which is the (7,0) mode at a higher amplitude than in the photograph immediately above it [12].

Sound spectra from the 18-cm *qing*, resting on the silk cushion, are shown in Fig. 12.17. The upper spectrum is recorded when struck and the lower spectrum 0.5 s later. The partial radiated by the (4,0) mode has the largest amplitude. Spectra recorded with the *qing* freely suspended by rubber bands are quite similar, indicating that most of the mode damping is due to sound radiation and that the silk cushion does not damp the vibration of the *qing* very much [12].

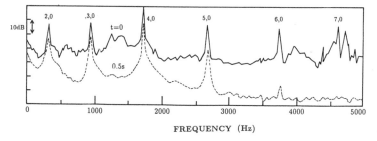

Fig. 12.17. Sound spectra from an 18-cm *qing* resting on a silk cushion [12].

Frequencies of the main partials in two *qing*, along with their ratios to the fundamental, are given in Table 12.2. Note that no harmonic relationship exists among the partials. The pitch of the *qing* appears to be determined by the fundamental (2,0) partial, although the strong (4,0) partial is clearly heard as an overtone.

Table 12.2. Mode frequencies and ratios to the fundamental in two *qing* [12]

Mode	18-cm *qing*		15-cm *qing*	
	$f_{m,n}$ (Hz)	$f_{m,n}/f_{20}$	$f_{m,n}$ (Hz)	$f_{m,n}/f_{20}$
(2,0)	346	1.00	434	1.00
(3,0)	953	2.75	1180	2.72
(4,0)	1751	5.06	2130	4.91
(5,0)	2691	7.78	3267	7.53
(6,0)	3748	10.83	4496	10.36
(7,0)	4644	13.42	6182	14.24
(8,0)	6255	18.08		
(9,0)	7363	21.28		

12.9. Crotals

Crotals (not to be confused with crotales, which are discussed in section 9.8) are closed bells with a pellet inside that strikes the inside surface when it is shaken. Most crotals are made of metal (usually copper, bronze, or brass), although some are made of clay or even wood. They vary in shape from nearly spherical to highly elongated, as shown in Fig. 12.18. Crotals may be the oldest bells of all. The Chinese were known to carry crotals into battle during the Shang dynasty.

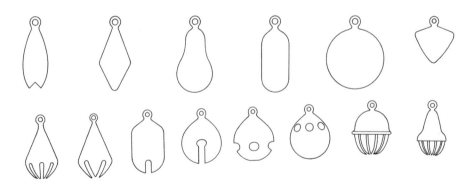

Fig. 12.18. Some shapes of crotals [after ref. 13].

In Central America, clay crotals that imitated natural pods were known between 2000 and 1000 BC [13]. By the eleventh century Inca craftsmen in South America had learned to cast crotals in silver and gold. Generally the crotal was an article of dress with great talismanic power as well as ornamental attraction. In the Yucatan, crotals were used as money. Necklaces found in southern Mexico have as many as 140 gold crotals.

Handbells

Crotals appear in the modern orchestra in the form of sleigh bells, hollow metal spheres up to 2 cm or so in diameter with a loose metal ball inside. They are often fastened in different sizes to a leather strap, and sometimes the strap is fasteneed to a wooden handle as shown in Fig. 12.19. Mozart used sleigh bells tuned in C, E, F, G, and A for the German Dances.

Fig. 12.19. Orchestral sleigh bells are a modern form of crotals.

12.10. Cow Bells

Cow bells have been used to track cattle in the Alps and other mountainous regions for centuries. The best made Swiss treichels or cow bells are musical instruments with a fine sound. Vibrational modes of a large 20-cm Swiss treichel are shown in Fig. 12.20.

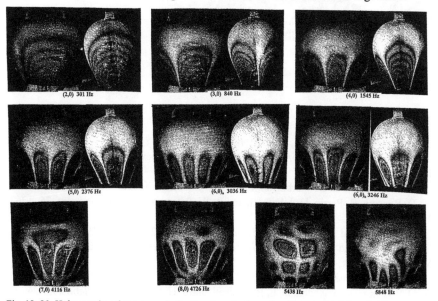

Fig. 12.20. Hologram interferograms of vibrational mode shapes in a Swiss treichel (cow bell).

Instruments used in orchestras and jazz bands are somewhat different from treichel both in appearance and sound. The sound is "drier," less bell-like, and the clappers are omitted. A set of chromatic cowbells is shown in Fig. 12.21, along with the straight-sided type generally used in jazz ensembles. Cowbells are used in Mahler's 6th and 7th symphonies, presumably to suggest distant cattle.

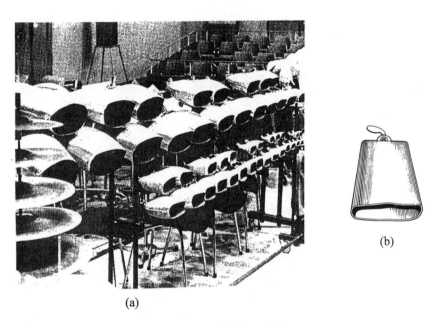

Fig. 12. 21. (a) Chromatic cowbells; (b) straight-sided jazz cowbell.

References

1. T. D. Rossing and R. Perrin, *Applied Acoustics* **20**, 41-70 (1987).
2. Lord Rayleigh, *The Theory of Sound*, Vol 1 (Macmillan, New York, 1894). (Reprinted by Dover, NY, 1945).
3. R. N. Arnold and G. B. Warburton, *Proc. Roy. Soc. London* **A197**, 238-256 (1949).
4. T. D. Rossing, R. Perrin, H. J. Sathoff, and R. W. Peterson, *J. Acoust. Soc. Am.* **76**, 1263-1267 (1984).
5. T. D. Rossing and H. J. Sathoff, *J. Acoust. Soc. Am.* **73**, 2225-2226 (1980).
6. N. H. Fletcher and T. D. Rossing, *Physics of Musical Instruments,* 2nd ed. (Springer-Verlag, New York, 1998).
7. T. D. Rossing, *Overtones* **27**(1), 4-10, 27 (1981).
8. T. D. Rossing, D. Gangadharan, E. R. Mansell, and J. H. Malta, *MRS Bulletin* **20**(3), 40-43 (1995).

9. T. D. Rossing, D. A. Russell, and D. E. Brown, *American J. Physics* **60,** 620-626 (1992).
10. T. D. Rossing and G. H. Canfield, Paper 4pMU11, Acoustical Society of America, Ottawa, 1993.
11. A. R. Trasher, "Qing" in *New Grove Dictionary of Musical Instrument* **3**, ed. S. Sadie (Macmillan, London, 1984).
12. T. D. Rossing and J. Tsai, *Acoustics Australia* **19**, 73-74 (1991).
13. P. Price, *Bells and Man* (Oxford University Press, Oxford, 1983).

Chapter 13
Eastern Bells

We began both of the previous chapters by pointing out that bells existed in China several centuries BC. I tell my students that to see the oldest of anything, go to China; to see the largest of anything, go to Russia. That is certainly true of bells.

The earliest known Chinese bells are small clapper bells of clay and copper from the third millennium BC. Beginning with the early Bronze Age (ca. 2000 BC), however, bronze became the material of choice for cast bells, as it remains to this day. The art of casting bronze developed to a high level of sophistication during the Shang dynasty (1766-1123 BC). Most bells were cast using the "piece-mold" technique, which began with a core model of the intended bell in clay. This model served as the imprint for the outer molds, which were produced by applying a layer of clay to the model surface. The outer molds were removed from the model in sections and fired. After firing, the outside of the core model was scraped down to leave a gap of the desired bell thickness between the core and the outer mold. The piece-mold technique was particularly convenient for casting oval or almond-shaped bells, which became the bells of choice, probably because of their two-tone behavior [1,2].

13.1. Ancient Chinese Two-Tone Bells

The earliest bronze bells, with clappers, were of two types: *ling*, which were suspended by means of a u-shaped loop and *duo* which were fitted with a handle. (The term *ling* is also used in some writings to include all bells). *Ling* go back as far as the beginning of the bronze age and might be considered the ancestors of all other bells.

More important in a musical sense, however, are the clapper-less two-tone bells. Although most ancient Chinese musical bells are similar in overall shape, they may be distinguished in four main classes:

nao, with arched rims and large tubular shanks which supported them in the mouth-upward position;

yongzhong, similar in shape to nao but with a ring or loop affixed to the shank so that they could be hung downward, slightly tilted toward the player;

niuzhong, identical in body shape to yongzhong but suspended vertically from simple u-shaped loops;

bo, with level rims and sometimes elaborate suspension devices from which they hang vertically.

Examples of these four classes are shown in Fig. 13.1.

Eastern Bells 165

Fig. 13.1. Examples of four classes of ancient Chinese bells: a) *nao*; b) *yongzhong*; c) *niuzhong*; d) *bo*.

Ancient Chinese bells were richly decorated. One of the most notable decorative features are the *mei* or nipples, large or small, generally arranged into 4 groups of 9, as can be seen in Fig. 13.1. Within each group of 9 are 3 rows of 3 *mei* each, separated by *zhuan* or ridges. The *gu* or striking area is often richly decorated, and Chinese characters are sometimes engraved in the area between groups of *mei*. Parts of a typical bell are identified in Fig. 13.2.

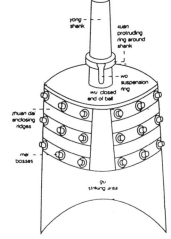

Fig. 13.2. Parts of a traditional Chinese two-tone bell.

The art of designing and casting ancient bells reached its peak during the Western Zhou (1122-771BC) and Eastern Zhou (770-249 BC) dynasties. Bell chimes occupied a prominent position in ancient Chinese ritual orchestras. Bells became status-defining objects. It is hardly any wonder that persons of wealth and power coveted bell chimes for their burial tombs.

Most musical bells were manufactured in scaled sets of chimes. They were probably used in sorts of rituals, as well as in musical performance. Chimes of bells were frequently buried in the tombs of royalty and noblemen, and that is how large numbers of ancient bells have been well preserved through the ages. The most remarkable set of bells discovered to date is the 65-bell set discovered in the tomb of Zeng Hou Yi (Marquis Yi of Zeng) in 1987. The richly-inscribed bells survived in excellent condition since about 433 BC due to the tomb filling with water. Also found in the tomb were bamboo flutes, panpipes (*xiao*), mouth organs (*sheng*), plucked string instruments (*qin*), drums, and stone chimes.

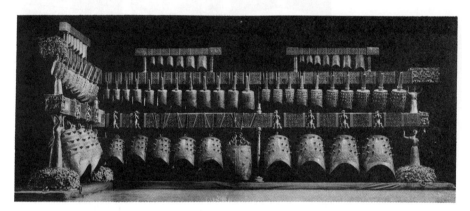

Fig. 13.3. Chime bells from the tomb of Zeng Hou Yi in Hubei province.

Study of the Zeng bells and many other sets of bells has provided musicologists with a great deal of understanding about ancient Chinese customs of musical theory and performance. The richly inscribed Zeng bells provided information about their intended pitch in terms of ancient pitch standards or *lü*. The musical system documented by the Zeng inscriptions gives the names of twelve moveable notes or *yin*, not unlike the do, re, mi of Western tradition. To the four simple *yin* names, *gong, shang, zhi, and yu,* are added suffixes such as *jue*, which raises a note by a major third and *zeng*, which raises it by two major thirds [2].

13.2. Vibrational Modes of Ancient Two-Tone Bells

In a round church bell or handbell, as we have discussed in Chapters 11 and 12, it is customary to denote a normal mode of vibration by the number of complete nodal meridians m and the number of nodal circles n. In a church bell, the principle modes (m,n) are given names such as *hum* (2,0), *prime* (2,1), *third* (3,1), etc., and they are carefully tuned to the desired frequencies. Similar normal modes are observed in almond-shaped Chinese bells, but because of their elongated shape, each mode becomes a doublet, which we designate as $(m,n)_a$ or $(m,n)_b$, as shown in Fig. 12.4a. The B-mode has a node at the spine or *xian*, while the A-mode has an anti-node at this location. Striking the bell at the center of the broad face will preferentially excite modes having an anti-node there, such as the $(2,0)_a$, $(3,0)_b$, etc., while striking it about halfway toward the *xian* or spine will preferentially excite modes such as the $(2,0)_b$ and $(3,0)_a$.

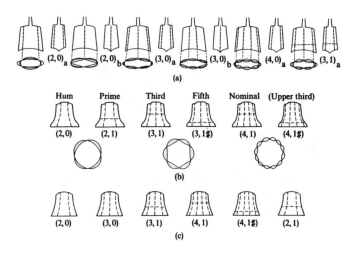

Fig. 13.4. First six modes of vibration in: a) Chinese two-tone bell; b) Western church bell: c) tuned handbell. Dashed lines indicate locations of nodes [3].

Holographic interferograms of 15 mode pairs in a modern Chinese two-tone bell are shown in Fig. 13.5. The shape of this bell, from the Suzhou Musical Instrument Factory, is based on ancient bells. The ratios of the B-mode to the A-mode frequencies range from 0.89 to 1.19 for the various mode pairs. In most cases, the B-mode has a higher frequency than the A-mode, as in other ancient Chinese bells studied [5]. The largest splittings occur in the (2,0) and (4,1) mode pairs, although no consistent pattern of mode-splitting size is noted [1]. The thickness of the bell varies from about 9 mm at the center of the faces to 15 mm near the ends. On the inner surface of the bell are six grooves, each approximately 10 mm wide and 10 mm deep. Similar grooves, sometimes found in ancient bells, can have a substantial effect on the mode doublet splitting (and hence on the interval between the A- and B-tones). The fundamental $(2,0)_b$ and $(2,0)_a$ modes, which largely determine the pitches of the two tones are tuned near to a minor-third interval (frequency ratio 1.20).

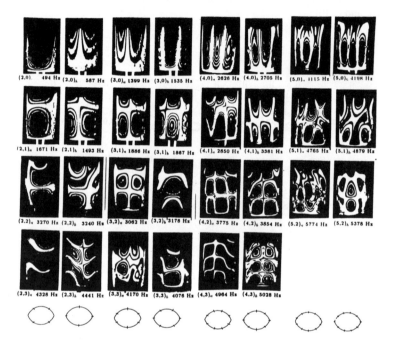

Fig. 13.5. Holographic interferograms of 15 mode pairs in a modern Chinese two-tone bell [5].

13.3. Interval Between the Two Tones

The inscriptions on the Zeng Hou Yi bells indicate two intended intervals between the *A*-tone and *B*-tone, corresponding to a major third and a minor third in Western notation. In fact, the interval was difficult for ancient bell founders to control to a high degree of accuracy. The measured intervals for the 65 bells, indicated in Fig. 13.6, show some clustering around 300 cents (minor third) and 400 cents (major third), but they range from about 210 to 490 cents.

Eastern Bells

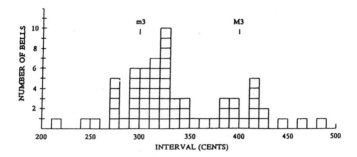

Fig. 13.6. Histograms showing the distribution of A-tone/B-tone intervals in Zeng Hou Yi bells. Intervals are expressed in cents (1/100 of a semitone on the scale of equal temperament or 1/1200 of an octave) [1].

An important question is how the ancient Chinese bell casters controlled the interval between the two tones. The following four methods might have been used:

1. By a suitable choice of the width-to-depth ratio, or eccentricity, of the bell;
2. In *yongzhong* and *niuzhong,* by controlling the height of the arch (*yu*) at the rim;
3. By varying the thickness of different parts of the bell wall during casting;
4. By varying the thickness of different parts of the bell wall after casting (i.e., cutting grooves or troughs in the cold metal).

A priori, one might expect the first method to have been the most convenient. Frequency ratios of mode pairs in almond-shaped cylinders with different width-to-depth ratios are shown in Fig. 13.7. Frequency splitting ratios in both the (2,0) mode (solid curve) and the (3,0) mode (dashed curve) increase with eccentricity, as expected. This has been further confirmed by finite-element calculations on bell shapes [6].

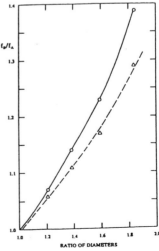

Fig. 13.7. Frequency splitting ratios of mode pairs in almond-shaped cylinders with different width-to-depth ratios. Solid curve is for the (2,0) mode, dashed curve for the (3,0) mode [7].

The mode intervals in the Zeng Hou Yi bells are shown as a function of the width-to-depth ratio in Fig. 13.8. In the middle row of bells, the measured intervals are quite close to the intended intervals (1.20 or 1.25, respectively, for minor or major third bells). However, there is little or no correlation with the width-to-depth ratios, which range from 1.29 to 1.37 for both groups of bells. In the upper row (filled circles) and bottom row (open circles and triangles), there is not only a lack of correlation between interval and width-to-depth ratio, but the measured intervals do not follow the intended intervals very well either. Note that there are no intended major-third bells in the upper row.

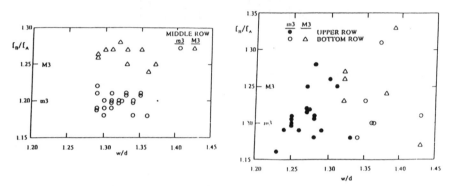

Fig. 13.8. Interval between A- and B- tones as a function of length-to-depth ratio in the Zeng Hou Yi bells: a) bells in the middle row; b) bells in the upper and bottom rows [1].

Other sets of chime bells show similar behavior. Intervals between the A-tone and the B-tone for all published bells are shown in Fig. 13.9. There is little, if any, correlation with width-to-depth ratio. Apparently the bell founders did not depend upon the width-to-depth ratio or eccentricity to control the interval between tones.

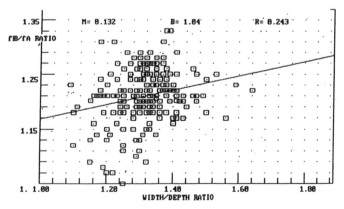

Fig. 13.9. B-tone to A-tone intervals for 205 ancient Chinese bells [1].

How about the height of the yu (mouth cutup) and the thickness? Our studies of over 200 bells for which physical measurements are available failed to show any real correlation between these parameters and the A-tone/B-tone interval either [8]. Did they rely on fine tuning after casting, perhaps by selective thinning or by cutting grooves in the bell? It is difficult to determine, after all these years, whether the bells were carved after casting or whether tuning grooves were added. Relatively few bells have been carefully examined on the inside, and corrosion over the years has obscured surface details.

Having no electronic instruments or reliable tuning standards, the ancient bell founders could tune only by ear. It is not at all unlikely that they cast many bells, selected the ones whose sounds they liked and melted down the others to be re-cast. Tuning of the A-tone may have been a more important selection criterion than the A-tone/B-tone interval. In the case of bells that were inscribed with a note name at the time of casting, the match to this designated note name would become an important criterion for acceptance, although the match in many bells was far from perfect, as we have noted. However well the ancient Chinese bell founders were able to predict or control the tone separation in the two-tone bells they cast, their casting skills were remarkable for their time. After thousands of years, many of the bells still remain musical instruments of high quality!

13.4. Temple Bells in China

Although two-tone chime bells continued to be cast during the Qin dynasty (221-208 BC) and even into the Han dynasty (206 BC-220 AD), their importance rapidly waned. New musical styles, in which bells played little or no role, were gaining favor. The few two-tone chime bells remaining from this period appear to be of a quality inferior to the bells from earlier periods. At the same time, expansion of the empire put the Chinese into close contact with cultures of central Asia, including India. With the spread of Buddhism from the third century, a new type of large round temple bell, known in China as *fanzhong*, developed in East Asia.

Medieval East Asian bells are generally cylindrical or barrel-shaped with a circular cross-section and a rounded top. They are considerably larger and thicker than the earlier chime bells, and wall thickness is more or less constant throughout the bell. Several temple bells are shown in Fig. 13.10. The rims of most early temple bells are mostly plane, but some Chinese temple bells have scalloped rims, rims with notches, or small protuberances that suggest tiny feet. The arched suspension device or *fanzhong* often takes the shape of a bent dragon. Their surface is divided into rectangular fields by a banded ornament that has been likened to the belt of a Buddhist priest's robe, and sometimes the strike points are marked by lotus-flower ornaments [9].

Fig. 13.10. Temple bells.

Temple bells, such as *fanzhong* in China and similar bells in Korea (*pomjong*), Japan (*bonshō*), Burma, and Tibet, are generally considered to be patterned after Indian prototypes, but few, if any, bells of this age remain in India, so this is still an open question. In monasteries, both Buddhist and Taoist, *fanzhong* came to be used for both religious and secular purposes, such as signaling time of day. Successive waves of anti-Buddhist persecution resulted in melting down of temple bells, as well as bronze Buddha statues, and so few early bells remain. One of the oldest remaining *fanzhong*, dating from 545 AD, is now in the Nara museum in Japan. From the Tang dynasty (618-907 AD) onward, large bells were placed, together with drums, in towers at the center of every city to announce the beginning and the end of the day.

The casting of large temple bells in China reached its zenith during the Ming dynasty (1368-1620). The largest such bell, shown in Fig. 13.11, was cast during the reign of the emperor Yong-le (1403-1424). Its bronze body, 4.5 m high with a maximum diameter of 3.3 m, is inscribed with 227,000 characters of Buddhist sutra text; its mass is estimated to be 52,000 kilograms. It is displayed in the "Temple of the great Bell" in Beijing.

Fig. 13.11. Yong-le bell (15th century in Beijing.

Mode frequencies of the Yong-le bell are shown in Fig. 13.12. Modes are designated by the numbers of nodal meridians and nodal circles, as in other bells we have discussed in this and the previous two chapters. The lowest mode of vibration in this large bell occurs at the very low frequency of 22 Hz [10].

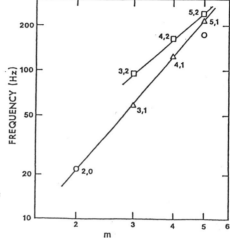

Fig. 13.12. Mode frequencies of the Yong-le bell as a function of the number of nodal meridians m. Each mode is labeled with the number of nodal meridians and the number of nodal circles (data from [10]).

From about the 12th century onward, many large bells in China were cast of iron, a much less expensive alternative to bronze. The most sonorous iron bells appear to have been of white cast iron rather than the more common grey cast iron [11]. The shapes and decorations of iron bells were quite similar to bronze bells. An iron bell is shown in Fig. 13.13.

Fig. 13.13. Iron bell.

Iron bells tend to be tonally inferior to bronze bells, largely due to the more rapid rate of sound decay. The faster sound decay with iron or steel bells is often attributed to greater internal loss in the material, but this is not necessarily the case. A more important factor is the greater efficiency of sound radiation (and thus the greater loss rate of vibrational energy) due to the higher sound velocity in iron or steel. Due to this greater radiation efficiency, the sound of a steel bell is initially louder than a comparable bronze bell, but the sound dies away more rapidly.

13.5. Korean Bells

Large temple bells have been cast in Korea for more than 1200 years. The most famous bell in Korea is the magnificent King Songdok bell cast during the Silla dynasty (771). Standing 3.66 m (12 ft) high, it has a mass of nearly 20,000 kg. The body has a flat protuberance called the *dang jwa* on which it is struck by a log suspended on ropes, and a chimney called the *eumtong* opens through the top, as shown in Fig. 13.12.

Fig. 13.12. King Songdok bell in Kyongju, Korea. Standing 3.66 m high, it has a mass of nearly 20,000 kg.

Figure 13.13 compares the profiles of typical Korean, Chinese, and Japanese temple bells. Note that the Korean bell, typical of bells cast during the Silla dynasty, reaches its greatest diameter at the point where it is struck, at about 1/3 of its height. The same Korean bell profile is compared to later Korean bells in Fig. 13.14.

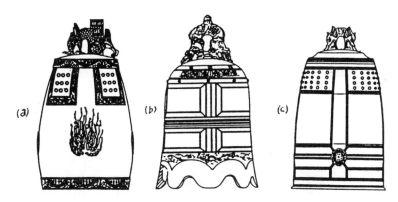

Fig. 13.13. Buddhist temple bells of: (a) Korea; (b) China; (d) Japan

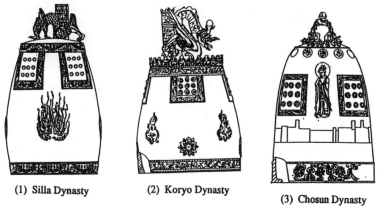

(1) Silla Dynasty   (2) Koryo Dynasty   (3) Chosun Dynasty

Fig. 13.14. Korean bells of three periods.

Although Korean temple bells are round, the mass of the *dang jwa* and other decorations that modify its symmetry are sufficient to create mode doublets, with one component having a node at the *dang jwa* and the other an antinode. The frequency separation of the two components is small, and this creates a slow beating or warble, which is considered to be a desirable characteristic of these bells.

The *eumtong* or chimney is said to act as a low-pass acoustic filter, favoring (in the case of the King Songdok bell) frequencies below 300 Hz [11]. Another acoustical feature of many large Korean bells is the circular depression in the ground or pavement beneath them, that, together with the volume of the bell and the chimney at the top, creates a Helmholtz resonator. In the case of the King Songdok bell, the depression, which can be seen in Fig. 13.12, is 94 cm deep .

A profile view of the King Songdok bell is shown in Fig. 13.15(a). Fig. 13.15(b) shows the *eumtong* or chimney at the top, and Fig. 13.15(c) shows one of the two arabesques that decorate the bell. The King Songdok bell has become a national treasure and symbol in Korea. The following is a translation of a portion of the inscription on the bell: "For its clear sound that reverberates across the country to teach without preaching and to lead us to the enlightenment of absolute truth, may the king and his royal house prosper forever."

Fig. 13.15. (a) Profile view of the King Songdok bell; (b) the *eumtong* or chimney at the top; (c) one of the two arabesques that decorate the bell.

A number of acoustical measurements have been made on the great King Songdok bell. A team led by Kim Yang-han used an array of 30 microphones (spaced 15 cm apart) for acoustic holography of the radiated sound field, and attached 9 accelerometers to measure the surface vibration. Strong mode doublets were found at 64.0625/64.375 Hz and 168.3125/168.4375 Hz. Mode shapes of the latter mode pair are shown in Fig. 13.16.

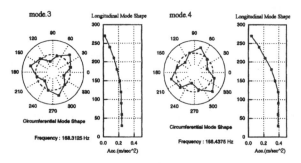

Fig. 13.16. Mode shapes for a mode doublet observed in the King Sondok bell by attaching 9 accelerometers [13].

Eastern Bells

In our laboratory we have studied the vibrational modes of a small Korean bell having a maximum diameter of 20 cm (19 cm at the mouth) and a height of 25 cm (excluding the *eumtong*). The thickness ranges from 11 mm near the middle of the bell to 8.8 mm at the rim. Holographic interferograms illustrating mode shapes in the bell are shown in Fig. 13.17. The modes are labeled $(m,n)_{a,b}$, where m gives the number of complete nodal meridians and n the number of nodal circles, as in Western church bells. The two components of each mode doublet are labeled by a and b; the b component has a node at the *dang jwa*, while the a component has an antinode at that location. In most cases, the a component is slightly lower in frequency due to the effect of the additional mass at the antinode [14].

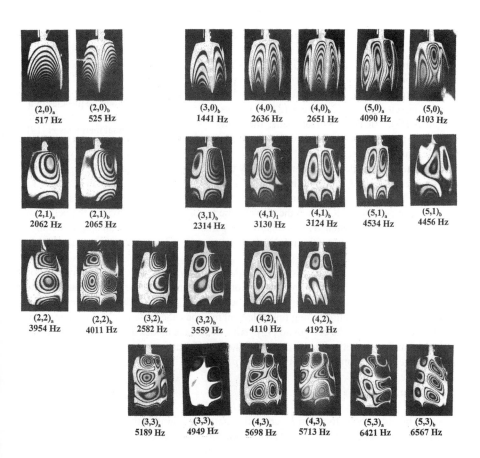

Fig. 13.17. Holographic interferograms showing mode shapes in a small Korean bell [14].

Mode frequencies are shown as a function of the number of nodal meridians m in Fig. 13.18. The three curves represent families of modes with 0, 1, and 2 nodal circles. Comparing these curves to similar graphs for church bells, carillon bells, and handbells, we note that there is a more complete family of $(m,0)$ modes than found in any of the above bell types. In church bells and carillon bells, only the $(2,0)$ ("hum") mode is devoid of nodal circles; handbells have $(2,0)$ and $(3,0)$ modes, and large bells may have $(4,0)$ and $(5,0)$ modes as well. The Korean bell does not have $(m,1^{\#})$ modes with nodal circles just above the mouth of the bell.

When the bell in Fig. 13.17 is struck to the side of the *dang jwa*, it has a warble whose frequency corresponds to the difference between the frequencies of the $(2,0)_a$ and $(2,0)_b$ modes (about 8 Hz). In large Korean bells, the slow beating (typically at about 1 Hz) is considered to be a desirable characteristic.

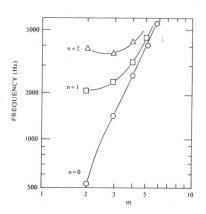

Fig. 13.18. Mode frequencies of a small Korean bell as a function of number of nodal meridians m [14].

The $(2,0)$ mode has the longest decay time (nearly 30 s), while the $(3,0)$, $(4,0)$, and $(2,1)$ modes have decay times in the range of 1-4 s. All other modes decay in less than 1 s. The long-lived $(2,0)$ mode radiates the fundamental partial which coincides with the perceived pitch of the bell. When the microphone is positioned on the axis of the bell a few centimeters from the mouth, a second-harmonic partial is observed with an amplitude about 20 dB below the fundamental. This is due to a second-order radiation process noted for handbells in section 12.2.

The dynamic characteristics of a large Yi-dynasty bell were analyzed both theoretically and experimentally by Chung, Kong, and Yum [15]. They noted that the $(2,0)$, $(3,0)$, and $(4,0)$ modes have a long decay time, and thus give Korean bells their lingering sound. Through calculations using finite element methods, they determined the effect of two and four *dang hwas* on the bell sound [15]. Kim, et al. have modeled a Korean bell structure as a ring-stiffened shell, and the results of their model agree quite well with vibrational modes observed in Korean bells [16].

13.6. Japanese Bells

The oldest bells found in Japan, called *dotaku*, belong to the Yayoi culture, which flourished from about 250 BC to 250 AD. Their form is an oval conoid, about 1½ times as high as wide, with a flange that extends up from the mouth and over the top, as shown in Fig. 13.19. Clappers were found buried inside some early bells and hammers were found buried with some later ones. In the fifth century Confucianism spread to Japan, and chimes of tuned bells (typically twelve) were introduced into ritual music for several great sacrifices performed by the emperor [17].

Fig. 13.19. A *dokatu* of the late Yayoi period in the Imperial Museum, Tokyo.

Fig. 13.20. The oldest temple bell in Japan, now hanging in the Myoshinji temple in Kyoto.

The introduction of Buddhism into Japan in the sixth century led to the development of the *sho*. The oldest such bell remaining in Japan is the Ojikicho bell (so called after one of the standard pitches with which it harmonizes), cast on Kyushu in 698 and now housed in the Myoshinji temple in Kyoto (see Fig. 13.20). It is still struck regularly today (every evening at sunset), thus being the oldest bell in the world in continuous usage. It measures 124 cm in height and 87 cm in diameter, and it has a fundamental frequency of 129 Hz. Altogether, 489 Japanese temple bells pre-dating 1615 are preserved today.

Large temple bells in the time of the Ojikicho bell came to be called *bonshō* or Brahmin bells. *Bonshō* are stuck at the "lotus" or striking point by a wooden log suspended from the ceiling. The resulting sound is considered to be the essence of Buddha. The largest temple bell was cast in 732 for the Todaiji temple in Nara. Standing 4.12 m high, 2.77 m in diameter and weighing over 26 tons, it was the largest bell in the world when cast.

A smaller *shō*, called *denshō*, was often used for signaling. Inside the temple it signaled the times of services and other happenings. In the homes of the wealthy, it was used to call guests to a tea ceremony, and in the kabuki theatre it was used for off-stage musical effects. One of the oldest *denshō*, 46 cm in diameter and dated 1217, hangs in the Koryuji temple in Kyoto.

13.7. Other Asian Bells

Although bells may have existed in India as long as in China, very little is known about the earliest bells. Hindu deities are often portrayed as carrying *singita* (crotals) or *ghanti* (open-mouth bells). Bells were sometimes hung on sacred animals, such as the cow and buffalo. A favorite place for wind bells was over the relics of saints buried in mounds or in *stupas*. Although Indian craftsmen did not produce enormous bells, they made some of the smallest and most delicate bells found anywhere.

In Indonesia, Hindu bells are found in seventeenth and eighteenth-century reliefs at Borobudur on the island of Java. The bells of Hinduism were silenced after the spread of Islam in the sixteenth century. Hindu priests and their bells were driven to the island of Bali.

Many of the early people of Central Asia were nomadic, and so small bells that could be carried from place to place were much more common than large bells. Marco Polo tells of bells carried by foot messengers, and bells were often attached to animals in order to keep track of them. Lamas in Tibet use handbells to attract the attention of the gods.

The Burmese have made some of the largest bells in Asia; three enormous ones hang in Burma today. In 1791 King Bawdawpaya had a large 88-ton bell cast at Mingun near Mandalay. With a diameter of 4.95 m and a height of 3.66 m, it is the largest soundable bell in the world. Another bell of 42 tons was cast in 1840 for the Shwe Dagon in Rangoon, as was a smaller bell of 16 tons. The great bell of Mingun has lived a life full of adventure. While being floated down the Irrawadi on its way to Calcutta, the barge carrying it capsized. The bell fell into the river but was retrieved and re-hung. In 1938 an earthquake caused it to fall, but once again it was raised and hung in a splendid new pavilion [17].

References

1. L. von Falkenhausen and T. D. Rossing "Acoustical and Musical Studies on the Sackler Bells" in *Eastern Zhou Ritual Bronzes from the Arthur M. Sackler Collections*, vol. 3, ed J. So (Abrams, New York, 1995).
2. L. von Falkenhausen, *Suspended Music* (Univ. California Press, Berkeley, 1993).
3. T. D. Rossing, D. S. Hampton, B. E. Richardson, H. J. Sathoff, and A. Lehr, *J. Acoust. Soc. Am.* **83**, 369-373 (1988).
4. T. D. Rossing and H-F Zhou, "Sound spectra of ancient Chinese bells in the Shanghai Museum," 117th meeting, Acoust. Soc. Am. (1989).
5. J. Tsai, Z. Jiang, and T. D. Rossing, *Acoust. Australia* **20**(1), 17-19 (1992).
6. A. Lehr, *Bericht uit Het Nationaal Beiaardmuseum* **16**, 14-   (1997).
7. U. J. Hansen and T. D. Rossing, *Proc. SMAC93* (Royal Swedish Academy of Music, Stockholm, 1994) 315-317.
8. T. D. Rossing and L. von Falkenhausen, *Proc. SMAC93* (Royal Swedish Academy of Music, Stockholm, 1994) 331-337.
9. L. von Falkenhausen and T. D. Rossing, *Proc. International Conf. on "Chinese Archeology Enters the Twenty-First Century"* (Xexue Chubanshe, Beijing, 1998) 407-434.
10. Chen T. and Zheng D., *Chinese J. Acoustics* **5**, 375-   (1986).

11. W. Rostoker, B. Bronson, and J. Dvorak, *Technology and Culture* **25**(4) 750 (1984).
12. Lee Byung-ho, *Collection of Papers on the Divine Bell of King Songdok* (Kyongju National Museum, Kyongju, Korea, 1999), 359-399.
13. Kim Yang-han, *Collection of Papers on the Divine Bell of King Songdok* (Kyongju National Museum, Kyongju, Korea, 1999), p.341-357.
14. T. D. Rossing and A. Perrier, *J. Acoust. Soc. Am.* **94**, 2431-2433 (1993).
15. S. C. Chung, C. D. Kong and Y. H. Yum, *KSME Journal* **1**, 133-139 (1987).
16. S. H. Kim, W. Soedel, and J. M. Lee, *J. Sound and Vibration* **173**, 517-536 (1994).
17. P. Price, *Bells and Man* (Oxford University Press, Oxford, 1983).

Chapter 14
Glass Musical Instruments

Glass musical instruments are probably as old as glassmaking. The 14th century Chinese *shui chan* consisted of nine glass cups struck by a stick. There are also 15th century Arabic references to musical cups (*kizam*) and jars *khaurabi*). Figure 14.1, from Gafori's *Theoria Musicae* (Milano, 1492), is a woodcut showing the musical use of glasses in a "Pythagorean" experiment.

In France, an instrument known as the "verillons" consisted of 18 glasses mounted on a board. The player tapped the glasses with long sticks.

At least as early as the 17th century it was discovered that wine glasses, when rubbed with a wet finger, produced a musical tone. Harsdorfer's *Deliciae Physicomathematicae* (Nuremburg, 1677) has the following prescription: "To produce a merry wine-music, take eight glasses of equal form; put in the one a spoonful of wine, in the other two, in the third three, and so on. Then let eight persons, with fingers dipped in wine, at the same moment pass them over the brims of the glasses, and there will be heard a very merry wine-music, that the very ears will tingle." [1]

Fig. 14.1. Woodcut from Gafori's *Theoria Musicae* showing glasses in a "Pythagorean" experiment.

14.1. The Glass Harmonica

Wineglasses can be musical instruments. The dinner guest who playfully rubs a wetted finger around the rim of a wineglass to make it "sing" is joining Mozart, Gluck, Benjamin Franklin, and others who have constructed or performed on glass harmonicas.

Glass harmonicas are basically of two types. One type employs vertical wine glasses arranged so that the performer can rub more than one glass at a time. A collection of glasses played in this manner is sometimes called a *glass harp*. The other type, called the *armonica* by its inventor Benjamin Franklin, employs glass bowls or cups turned by a horizontal axle, so the performer need only touch the rims of the bowls as they rotate to set them into vibration (Franklin's instruments can be seen in several museums, including the Franklin Museum in Philadelphia). Both types of instruments are shown in Fig. 2.

Fig. 14.2. (a) Glass harp played by Jamey Turner; (b) Franklin-type armonica shown with its builder Gerhard Finkenbeiner.

The practice of rubbing the rims of glass vessels to produce steady tones became popular in Europe during the 18th century. In 1746 the composer Christoph Willibald Gluck gave a concert in London advertised as "a concert on 26 drinking glasses tuned with spring water, accompanied with the whole band, being a new instrument of his own invention, upon which he performs whatever may be done on a violin or harpsichord." [1] (A bit of an exaggeration, no doubt).

On a trip to England, Benjamin Franklin heard a set of glasses played by E. H. Daval in Cambridge. In a letter to a friend in Italy in 1762, Franklin wrote, "Being charmed by the sweetness of its tones, and the music he produced from it, I wished only to see the glasses disposed in a more convenient form, and brought together in a narrower compass, so as to admit of a greater number of tunes, and all within the reach of hand to a person sitting before the instrument" [2]. Thus, the armonica was born.

Famous Europeans who became armonica players included Marie Antoinette and Franz Mesmer, the hypnotist. Mozart learned to play the instrument and composed several pieces for it, as did Beethoven. Franklin himself composed several songs for the instrument, and Donizetti wrote a part for glass harmonica in the accompaniment to Lucia's mad scene in *Lucia di Lammermoor*, although the part is nowadays generally played on a flute.

Several attempts have been made to add a keyboard. Marianne Kirchgessner was reported to be a virtuoso on the keyboard version during the 19th century, and it may have been the keyboard version that Mozart first heard. Organist E. Power Biggs arranged to have the Schlicker organ company build a keyboard version, on which he played a concert in 1956 to commemorate the 250th anniversary of Franklin's birth (and the 200th anniversary of Mozart's birth), but apparently the instrument was not particularly successful.

In this century interest in the glass harmonica has been kept alive by Bruno Hoffman and other dedicated performers. In 1988, a group of glass music enthusiasts banded together to form an organization called Glass Music International, which has held several glass music festivals and now publishes a quarterly called *Glass Music World*.

14.2. Vibrational Modes of a Wineglass

The vibrational modes of a wineglass rather closely resemble the flexural modes of a large church bell (Chapter 11), a small handbell, or a Chinese *qing* (Chapter 12). The principal modes of vibration result from the propagation of bending waves around the glass, resulting in $2m$ nodes around the circumference.

In the lowest mode with $m=2$ (corresponding to the (2,0) mode in a bell), the rim of the glass changes from circular to elliptical twice per cycle, as shown in Fig. 14.3. To a first approximation, at least, the radial and tangential components of the motion are proportional to $m \sin m\theta$ and $\cos m\theta$, respectively; for the (2,0) mode the maximum tangential motion is half the maximum normal motion. This means the glass can be excited by applying either a tangential force (rubbing with a finger) or a radial force (with a violin bow or a mallet).

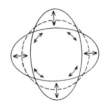

Fig. 14.3. Motion of the mouth of the glass in the (2,0) mode. The maximum tangential motion is half the maximum normal motion.

Several normal modes in a wineglass are shown in Fig. 14.4. At higher frequencies, the motion is concentrated mainly near the rim of the glass.

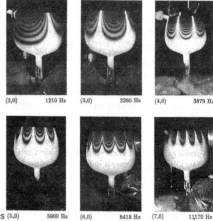

Fig. 14.4. Holographic interferograms of 6 normal modes of vibration in a wineglass [3].

Figure 14.5 (a) shows the modal frequencies in several wineglasses and brandy snifters as functions of m. The modal frequencies lie pretty well along straight lines with slopes of 2 on a logarithmic graph; that is, frequency is nearly proportional to m^2, as in a cylindrical shell. A similar plot in Fig. 4(b) shows modal frequencies for three armonica bowls blown by Gerhard Finkenbeiner, a scientific glassblower in Waltham, Massachusetts.

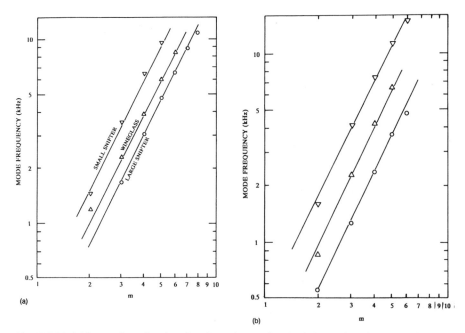

Fig. 14.5. Modal frequencies as function of mode number m, where $2m$ is the number of nodal meridians: (a) a wineglass and two brandy snifters; (b) three armonica bowls.

For any given mode, the modal frequency is roughly proportional to the thickness and inversely proportional to the square of the radius. Why the square of the radius? Because the speed of flexural waves is proportional to $\sqrt{f}$. Thus, if the time for a wave to travel the circumference is taken to be its period, we can say that $1/f = 2\pi r/k\sqrt{f}$, from which $f \propto 1/r^2$.

14.3. Rubbing, Bowing, Striking

Since the modes of vibration of glasses or bowls have both normal and tangential components, they can be excited by either a normal or a tangential force. Glass harmonicas are usually played by rubbing the rim of the glass or bowl tangentially with a wet finger, but striking the glass with a mallet or bowing it radially with a violin bow also sets it into vibration.

A moving finger excites vibrations in the glass through a "stick-slip" action, much as a moving violin bow excites a violin string. During a part of a vibration cycle, the rim of the glass at the point of contact moves with the finger; during the balance of the cycle it loses contact and "slips" back toward its equilibrium position. This results in a sound that consists of a fundamental plus a number of harmonic overtones, although not nearly so many as the sound of a violin. The location of the maximum motion follows the moving finger around the glass.

Although rubbing a glass with a wet finger tends to excite only the (2,0) mode and its harmonics, tapping it excites the other modes as well. By bowing it radially with a violin bow, we have been able to excite either the (2,0) mode or the (3,0) mode (although not at the same time), and higher modes may be possible as well.

Figure 14.6 shows sound spectra from a wineglass excited by rubbing with a wet finger, tapping with a yarn-wrapped vibraphone mallet, and bowing with a violin bow. In Fig. 14.6(d), the glass was touched with two fingers at locations that suppress the (2,0) mode and thus encourage the (3,0) mode. It should be noted that bowing is also a "stick-slip" process, and only one mode can couple in a stable way to the bow at the same time unless two mode frequencies are near enough to having a harmonic relationship that mode locking can occur.

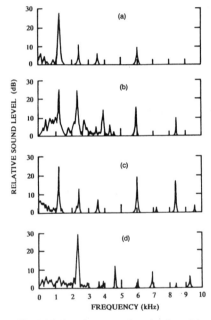

In the wineglass spectra in Fig. 14.6, the fifth harmonic of the (2,0) mode and the fundamental of the (5,0) mode are close together in frequency (5990 and 5997 Hz, respectively), and because of resonance coupling they appear rather prominently in all four spectra. [Coupling also appears between the seventh harmonic of the (2,0) mode and the fundamental of the (6,0) mode (8380 and 8423 Hz, respectively.]

Fig. 14.6. Sound spectra of a wineglass: (a) rubbing with a wet finger; (b) tapping with a yarn-wrapped mallet; (c) bowing with a violin bow; (d) bowing so as to excite the (3,0) mode.

Bowing with a violin bow [Figs. 14.6(c) and 14.6(d)] excites several modes, although the (2,0) mode is easily the strongest (even more than the sound spectra indicate, because the sound radiation efficiency increases rapidly with frequency near the coincidence frequency, and thus the higher-order modes radiate much more efficiently than the (2,0) mode). In Fig. 14.6(d) an effort was made to suppress the (2,0) mode and to enhance the higher modes by constraining the glass at appropriate points with the fingers.

Glass Musical Instruments

As the finger moves around the rim of the glass, the region of maximum vibration follows the moving finger, resulting in a sound that pulsates with about 4 to 8 beats per second, depending upon the speed of the player's finger, as shown in Fig. 14.7.

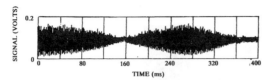

Fig. 14.7. Oscillograph of sound pressure from a large brandy snifter, showing a beat rate of 5 Hz (corresponding to 1¼ revolutions of the finger around the glass per second).

The player has relatively little control over the steady-state sound of a glass harmonica, as compared with other musical instruments. Moving the finger faster increases the level of the radiated sound, but only by 10 dB or less, unlike the violin where the player has considerable control over loudness by varying the bow speed. Furthermore, the player does not have the means to add higher harmonics, a feature that is associated with crescendos on most musical instruments.

14.4. Selecting and Tuning the Glasses

A glance at available photographs of glass harmonica performers suggests that a rather wide variety of glasses can be used. Brandy snifters appear to be quite widely used. Wineglasses with a large bowl diameter and a slightly narrower mouth appear to be popular as well. Makers of Franklin-type armonicas apparently prefer a harder quartz glass for durability. Glasses that have a clear and persistent ring when tapped will probably play well.

Finding glasses with similar qualities over a large range of sizes is a problem. One builder reported making over half the notes of a 3-octave instrument from 8-oz glasses by slicing off layers of varying size from the rim using a carbide tool and a torch [4]. Fine tuning can be accomplished by grinding. Grinding down the rim raises the frequency, whereas thinning the glass bowl near the base lowers the frequency. Fire polishing the rim raises the frequency slightly.

Fine tuning can also be accomplished by adding water (or wine), but the range of tuning is small. We found that filling various glasses about ¼ full lowered the playing frequencies from 0.3% to 0.9% (5 to 15 cents). Filling the same glasses half full lowered the frequencies up to 6% (one semitone or 100 cents). Water affects the playing quality, however, and during a long concert or rehearsal the liquid level can change because of evaporation. Thus it is better to fine-tune the glasses by careful grinding.

14.5. Verrophone

The verrophone uses tuned glass tubes rather than bowls. Two different verrophones are shown in Fig. 14. 8. The vertical tubes are played by rubbing in the same way as the wineglasses in a glass harp. The larger radiating surface can produce a greater sound output.

Fig. 14.8. The verrophone consists of tuned glass tubes that are played by rubbing similar to the glass harp. Two versions by different makers are shown.

14.6. Glass Bells

Glass is a hard material that vibrates with relatively low damping loss, and thus it is quite a good material for making small bells. Furthermore, it has quite a low melting temperature, and thus is easily shaped by standard glassblowing techniques. Although most glass bells are primarily decorative, the Sasaki Crystal company has produced tuned handbells similar in shape to the bronze handbells described in Chapter 12.

Gerhard Finkenbeiner, a scientific glassblower in Waltham, Massachusetts largely responsible for the resurgence of interest in the Franklin armonica (see Fig. 14.2b), developed a bell synthesizer that uses quartz fibers whose vibrations are picked up by a sensor and amplified to produce bell-like sounds. Several churches in New England have installed "tower bells" that make use of Finkenbeiner's synthesizer.

14.7. The Glass Orchestra

The Sasaki Crystal company has produced a wide variety of glass musical instruments, including glass marimbas, glass chimes, glass trumpets, glass horns, glass alpenhorns, and glass flutes. I had the privilege of hearing a number of these instruments played at an international musical acoustics meeting in Tokyo in 1992, and I have also enjoyed a videotaped performance by a "glass orchestra."

The Kassel Glass Orchestra in Germany, led by Walter Sons, plays on a variety of glass percussion and wind instruments, including vases, bowls, glass spheres, dishes, sheets of flat glass, glass flutes, and glass tubes [5].

In Lund Sweden, scientific glassblower Leif Lundberg has made instruments such as the trombone and tuba out of glass. His trombone is listed in the Guiness Book of Records (Swedish version) as the world's only playable trombone made entirely out of glass.

Several different glass companies make glass flutes

14.8. Glass Instruments of Harry Partch and Jean-Claude Chapuis

Composer-inventor Harry Partch constructed his own musical world of microtones, elastic octaves, and percussion instruments. "I have often dreamed of a private home with a wooden stairway which is in reality a Marimba Eroica," he writes, "with the longest block at the bottom and the shortest at the top. The owner could stipulate his favorite scale, then bounce up to bed at night hearing it." Partch's rich heritage of percussion instruments includes at least two glass instruments: "cloud chamber bowls" and "mazda marimba."

The tuned "cloud chamber" bowls, shown in Fig. 14.10, are actually cut from acid carboys, while the mazda marimba consists of light bulbs of various sizes.

Fig. 14.10. Harry Partch's "cloud chamber" bowls, which are really cut from acid carboys (photo by Steve Hockstein).

French composer and instrument maker Jean-Claude Chapuis has developed a number of glass instruments. His *glass balafon* is a marimba-like instrument with a set of cylindrical glass rods set over a box-like resonating chamber and played with mallets. The *crystallophone* is another mallet instrument with flat bars of plate glass. Several of these instruments are shown in Fig. 14.11.

Fig. 14.11. Jean-Claude Chapuis with several of his glass instruments, including a crystallophone, balafon, glass harp and glass armonica [6].

14.9. Other Glass Instruments

Quite a number of performers and inventors have experimented with glass instruments. Oliver Di Cicco's crystal harp is shown in Fig. 14.12.

Fig. 14.12. Oliver Di Cicco's crystal harp

References

1. D. A. Smith, "Acoustic Properties of the Glass Harmonica," Dept. Of Aerospace and Mechanical Sciences, Princeton Univ. (1973)
2. B. Franklin, letter to Giombatista Beccaria, Turin, Italy (1972). (Museum of Fine Arts, Boston)
3. T. D. Rossing, *J. Acoust. Soc. Am.* **95** (1994), 1106.
4. N. L. Rehme, *Glass Music World* **4**(2) (1990), 3.
5. E.-M. Stiegler, *Scott information* **55** (1991) 17.
6. B. Hopkin, *Gravikords, Whirlies & Pyrophones* (Ellipsis Arts, Roslyn, NY, 1996)
7. <www.mobiusmusic.com/chrystal.html>

Chapter 15
Other Percussion Instruments

In addition to the instruments we have described in the previous chapters, there are literally thousands of other percussion instruments, including many ethnic instruments found in various parts of the world, unusual instruments that find occasional use in bands and orchestras, and experimental percussion instruments. In this chapter we will describe a small sampling of these "other" instruments.

15.1. Anklung

The anklung is a bamboo instrument found in several areas of Indonesia but especially popular in West Java. Each instrument consists of two or three bamboo tubes, tuned in octaves, with protrusions that slide in a grooved bamboo frame. A tongue-shaped segment is cut out of one side of each tube, the size of the segment determining the frequency of vibration of the instrument. The instrument is generally held in the hands and shaken sideways by the player, although sometimes they are mounted in a frame so that one player can play the entire set, as shown in Fig. 15.1.

Anklung generally employ pentatonic tuning of the *pelog* or *slendro* type similar to gamelan instruments (see Fig. 7.7), although some sets are tuned diatonically. Each tube is tuned by shaping the tongue, while the uncut portion of the tube, which serves as a tubular resonator, is tuned by moving a plug to change its acoustic length.

Fig. 15.1. Anklung.

15.2. Deagan Organ Chimes

Although the J. C. Deagan Company was known mainly as a manufacturer of marimbas, xylophones, vibraharps, and chimes, they also marketed an extensive line of other instruments as well.

Their organ chimes or shaker chimes, shown in Fig. 15.2, appear to have been inspired by Balinesian angklungs but they are made of metal rather than bamboo. Each individual note consists of four specially-shaped tube chimes of nickel-plated metal. One chime is tuned to the fundamental pitch, one is tuned an octave higher, and two more are tuned two octaves above the fundamental. At the bottom of each closed tube are two metal tabs that glide back and forth in a short groove, as in the angklung, and strike wood pieces at either end of the groove.

The individual note frames are hung on a rack, as shown, with the naturals in the bottom row and sharps and flats in the top row. The player sounds the chimes by shaking the frames as they hang in the rack. The three-octave set in Fig. 15.2 has 37 frames. The largest set Deagan made was a four-octave set extending from G_2 to G_6 [1].

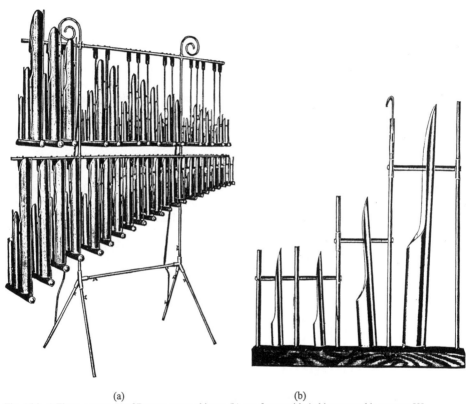

(a) (b)
Fig. 15.2. a) Three-octave set of Deagan organ chimes; (b) one frame with 4 chimes tuned in octaves.[2]

Tuning Deagan organ chimes is similar to tuning Choirchimes (section 12.7), angklung (section 15.1) or ordinary tuning forks. The vibrating tongue is tuned by removing mass at the end to raise its frequency or removing mass at the base to lower its frequency. Then the resonator is tuned by moving the stopper at the end of the tube until the frequency of the resonator matches that of the tongue, as shown in Fig. 15.3 [1].

Deagan also made aluminum chimes, which were similar to organ chimes except that they were made of aluminum rather than bronze and they have only three rather than four tubes.

15.3. Other Deagan Instruments

John Calhoun Deagan, an Irish clarinetist with a strong interest in science, emigrated from England to the United States in 1879 at age 27. He became interested in musical acoustics through reading Helmholtz' *On the Sensations of Tone as a Physiological Basis for the Theory of Music* and through hearing Helmholtz lecture in England. In 1880 he produced his first musical instrument, a set of "J. C. Deagan Musical Bells," a set of steel bars mounted in a wooden case. He applied his knowledge of physics to the tuning and mounting of the bars, which resulted in the first scientifically-designed mallet percussion instrument and lifted the glockenspiel (see section 7.1) to the status of an orchestral instrument. During his lifetime, Deagan probably invented more percussion instruments than any other person in history. In 1986 the Percussive Arts Society, in celebration of its 25th anniversary year, reprinted a collection of Deagan catalogues, which included 126 kinds of orchestra bells, 160 kinds of Deagan xylophones, 100 kinds of marimbaphones, and a complete line of Deagan "novelty" instruments such as organ chimes, aluminum harps, musical rattles, etc. [2]

Deagan's *aluminum harps* consisted of nickel-plated "bell-metal" tubes of various lengths mounted in an upright position on a sounding board to which they were attached by thumb screws. They are played by rubbing with resined gloves, similar to the more familiar solid stroke rods. Each tube acts as its own resonator, although they could not be tuned to the resonance frequency of the vibrating tubes, as far as I can see, because the speed of sound in air is much lower than the speed of longitudinal waves in brass or bronze. A vibrato could be obtained by moving a finger over the open end of the resonator (like the vibraharp or vibraphone, see section 7.3). A Deagan aluminum harp is shown in Fig. 15.3.

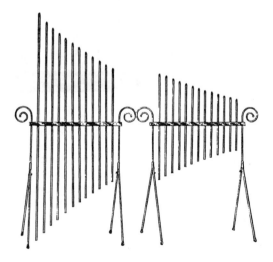

Fig. 15.3. Deagan "aluminum" harp consisting of tubes that are stroked with rosined gloves [2].

Other Percussion Instruments 195

Deagan's *musical rattles* consisted of "a heavy alloy steel bar permanently mounted in a frame in connection with a Deagan patent resonator." The bar was set in vibration by being struck by a clapper and the clapper was operated by a cog. Individual rattles were played by picking them up and whirling them. Altenatively, a series of rattles were mounted on a frame and operated by individual cranks "same as a coffee mill." [2]

The Deagan *musical coins* were toothed steel gears, 3 to 6 inches in diameter, that were played by spinning them on a marble top table. They were available in "any range or scale desired up to and including 2 octaves."[2]

15.4. Instruments of Harry Partch

Besides the glass instruments of Harry Partch (see section 14.7), we should mention a few other of his innovative percussion instruments. His *boos* (for bamboo marimbas) consists of 64 pieces of bamboo tuned by cutting tongues of the right lengths, as shown in Fig. 15.4. .

Fig. 15.4. Boos (bamboo marima) of Harry Partch (Photo by Steve Hockstein).

Fig. 15.5. Spoils of war includes a wood block, shell casings, strips of steel, and a gourd.

His *spoils of* war, shown in Fig. 15.4, consists of a variety of items: a Pernambuco (wood) block, seven brass shell casing, four "cloud chamber" bowls, two tongued pieces of bamboo with open ends, three "whang guns" (strips of spring steel controlled by pedals), and one gourd (which could be scraped to give a rasping sound).

15.5. Mark Tree

The *mark tree*, named after its creator Mark Stevens, consists of a series of small chime rods, graduated microtonally from low to high and suspended from a wood bar. The player glissandos up and down the scale with fingers or a triangle beater. The instrument was used for scene changes in the TV show "Charlie's Angels."

15.6. Instruments of Bernard and François Baschet

Bernard and François Baschet have built several unusual percussion instruments using familiar sources of vibration such as strings, rods, and bars. One example is the *cristal* shown in Fig. 15.6 (a). The basic vibrating system consists of a vertical glass rod attached to a horizontal threaded steel rod, as shown in Fig. 15.6 (b). The position of the glass road and the tuning weight are adjustable. The player excites the glass rod by stroking with wet fingers, and the glass rod sets the steel rod into vibration. The vibrations are transmitted to the sheet metal radiator through the steel gum attached to the threaded rod.

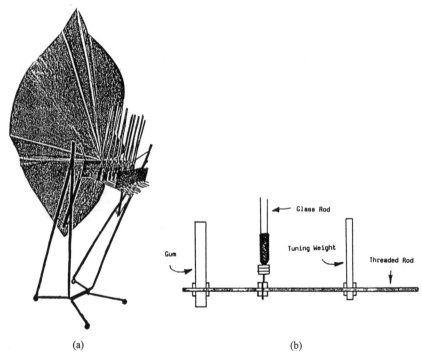

(a)   (b)
Fig. 15.6. (a) Cristal by B. and F. Baschet; (b) vibrating system in the cristal consisting of a vertical glass rod connected to a horizontal threaded steel rod. [4]

Several of the Baschet brothers' instruments use conical sheet metal resonators, sometimes mounted on the ends of thin metal rods so that they resemble flowers, as in the *nanny goat*. Others use piston pipe resonators, which consist of a pipe of moderately large diameter (up to 42 cm) closed by a rubber membrane. Two disks of plywood or aluminum reinforce the membrane on both sides, and these are excited by a wire or rod attached to the vibrator. Actually the piston pipes are designed to couple to several different vibrators, and there may be many pipes, tuned to different frequencies in an instrument.

Yet another resonator used in the Baschets' instruments are spring steel wires, which they call "whiskers." They are generally between 60 and 100 cm in length, and a large number of whiskers of varying length are attached to the body of an instrument. In addition to vibrating sympathetically, they bend and sway and occasionally contact one another. In addition to their acoustical role, the whiskers add an intriguing visual element. Both whiskers and cone resonators are seen in the *aluminum piano* in Fig. 15.7. [4].

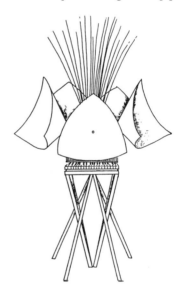

Fig. 15.7. Aluminum piano of B. and F. Baschet employs whiskers and cones as resonators.

15.7. Lithophones

Lithophones use vibrating stones to create sound. The ancient Chinese were fond of stone chimes, many of which have been found in ancient Chinese tombs. In a complete orchestra it was desirable to include instruments of stone, metal, silk, bamboo, wood, skin, and earth. A typical stone chime was shaped to have two arms of different lengths joined in an obtuse angle, as shown in Fig. 15.8. The stones were generally struck on their longer arm with a wooden mallet. Sometimes the stones were richly ornamented.

Several sets of stone chimes have been found in ancient Chinese tombs. A lithophone of 32 stone chimes in the tomb of the Marquis Yi (which also contained the magnificent set of 65 bells shown in Fig. 13.3) are scaled in size, although their dimensions do not appear to follow a strict scaling law [5].

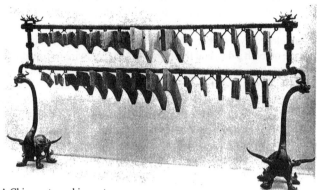

Fig. 15.8. A Chinese stone chime set.

Music historian Fritz Kuttner examined a set of 12 Chinese stone chimes in the Royal Ontario Museum (Toronto) and found that most of them sounded frequencies that were very close (within 4 cents) to notes on a Pythagorean scale, except for one chime which was close to a just major third above the tonic note and one chime that may have conformed to a "narrow fifth."[6]

In later times, the Chinese made stone chimes of jade. Holographic interferograms showing some of the modes of vibration of a small jade chime are included in Fig. 15.9.

15.9. Holographic interferograms of vibrational modes in a small jade chime.

Another jade object found in ancient Chinese tombs was the *bi* disk or *pi* disk, which may have been a type of lithophone. These disks have a variety of diameters, ranging from less than one inch up to 18 inches. At the center is a hole whose diameter ranges from 1/6 to 1/3 the outer diameter of the disk (jade disks with center holes larger than 1/3 the outer diameter are generally given the names *huan* or *yüan* instead). Kuttner [6] presents several arguments to support his theory that jade *pi* disks, such as those in Fig. 15.10 were used as musical percussion instruments.

Other Percussion Instruments   199

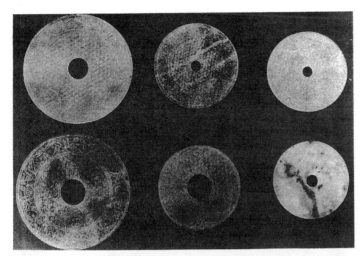

Fig. 15.10. Six jade *pi* disks from the tomb of Tou Wan (Western Han dynasty). [6]

Several old lithophones from Vietnam are in the Musée de l'homme in Paris. These ancient lithophones were clearly tuned by flaking and yield two sonorous pentatonic scales. They are 65 to 100 cm long, 10 to 15 cm wide, and between 5 and 11 kg in mass. [7]

Icelandic lithophones generally make use of basaltic, isotropic stones which, as a result of climatic changes, have split into thin slices or slabs. They are placed on two parallel strips of wood and played with small, hard mallets. [7]

Figure 15.11 shows the Till Family Rock Band with a lithophone probably made in the 1870s using stones found in the Lake District of northwestern England. The 50 stones, in two layers, were struck with wooden hammers covered with felt. The Tills gave concerts in the United States as well as throughout England. A smaller instrument was presented to the Metropolitan Museum of Art in New York by William Till.[8]

Fig. 15.11. Till family rock band with a lithophone probably made in the 1870s using stones from the Lake District of northwestern England (courtesy of Michael Till).

15.8. Ceramic Instruments of Ward Hartenstein

Musical instruments of clay have been known for many centuries, but none are more attractive or sonorous than those of contemporary sculptor/musician Ward Hartenstein. One of his first instruments was a ceramic tongue drum in which cantilevered bars, carved from a solid slab, vibrate over a bowl resonator [9]. From this developed the petal drum, shown in Fig. 15.12. Another innovative instrument is the bell tree shown in Fig. 15.13.

Fig. 15.12. Ceramic petal drum created by Ward Hartenstein.

Fig. 15.13. Ceramic bell tree by Ward Hartenstein.

In addition to creating ceramic instruments, Hartenstein performs on his instruments and composes music for them. His clay marimba, shown in Fig. 15.14, is a three-octave pentatonic instrument. The lower octave is slung directly underneath the upper two, permitting play from both sides as well as some unusual sticking patterns for four-mallet playing. Noting a lack of regard for visual components of musical performance in many percussion ensembles, he champions performances in which visual and aural components are on equal creative footing [9].

Fig. 15.14. Clay marimba of Ward Hartenstein, a three-octave pentatonic instrument with clay bars and resonator.

15.9. Thunder Sheet

Once used mainly for sound effects in the theatre, the thunder sheet has now become an orchestral instrument. The orchestral version is typically about 1.2 m (4 feet) wide and from 2 to 4 m (6 to 12 feet) long. It may be hung (from a step ladder, for example) and stuck with a soft beater, or it may be held by two players who shake it. A large thunder sheet is used in the Alpine Symphony of Richard Strauss. In his "First Construction," John Cage calls for five thunder sheets of different sizes.

15.10. Typewriter

When we think of the musical use of a typewriter, we probably think first of Leroy Anderson, who wrote an orchestral piece "The Typewriter" (1950). However, it is also scored in Erik Satie's "Parade" (1917), Ferde Grofé's "Tabloid" (1947) and Rolf Liebermann's "Concert des Changes." As word processors replace typewriters in offices and homes, perhaps typewriters will be preserved mainly as orchestral instruments.

References

1. B. Hopkin, *Experimental Musical Instruments* **9** (2), 9-16 (1993).
2. J. C. Deagan catalog H, *Percussive Notes* **24**(3/6), 172 (1986).
3. D. Drummond, *Zoomoozophone Primer*, <www.spyral.net/newband/zoomprimer.htm>.
4. B. Hopkin, *Experimental Musical Instruments* **3**(3), 4-10 (1987)
5. A. Lehr, *Acustica* **83** (1997), 320-336.
6. F. Kuttner, *The Archaeology of Music in Ancient China* (Paragon House, New York, 1990).
7. E. Davidsson, *Experimental Musical Instruments* (Sept. 1998), 29-31.
8. A. M. Till, *Experimental Musical Instruments* **7**(5), 12-13.
9. W. Hartstein, *Percussive Notes* **24** (4), 30-32 (1986).

Name Index

Achong, A. 127
Alexis, C. 108, 114, 115, 116, 125
Anderson, C. A. 20
Anderson, L. 202
Ando, S. 46
Arnold, R. N. 88, 162
Ayers, L. 75, 78
Banu, H. 145
Baschet, B. and F. 196
Bassett, I 35, 46
Beethoven, L. 183
Benade, A. H. 20
Bernoulli, D. 67
Bigelow, A. 143, 145
Biggs, E. P. 183
Blades, James 1, 4, 63, 106
Bork, I. 46, 58, 59, 63
Bossomaier, T. 106
Boverman, J. 127
Brindle, R. S. 10, 63, 78
Bronson, B. 181
Brown, D. E. 163
Cage, J. 202
Cahoon, D. E. 35, 46
Canfield, G. H. 163
Chaigne, A. 63, 106
Chapuis, J.-C. 190
Charnley, T. 144, 145
Chen T. 18;0
Chiaverina, C. J. 127
Chladni, E. F. F. 88, 91
Chitre, R. 127
Chung, S. C. 178, 181
Christian, R. S. 20
Colgrass, M. 37
Dagan, E. A. 46
Davidsson, E. 202
Davis, R. E. 12, 20
De, S. 19, 20
Deagan, J. C. 54, 74, 194

DeAlba, J. 127
Defrance, J. 127
DePont, J. 144
Deval, E. H. 183
DiCicco, O. 190
Doutaut, V. 63
Drummond, D. 76, 202
Dunlop, J. I. 71, 78
Dunn, H. K. 46
Dvorak, J. 181
Eggen, J. H. 144
Euler, L. 67
van Eyck, J. 128, 130
von Falkenhausen, L. 180
Ferreyra, E. 122, 123, 127
Finkenbeiner, G. 183, 185, 188
Fleischer, H. 12, 20
Fletcher, H. 35, 46
Fletcher, N. H. 20, 46, 51, 63, 78, 88, 106, 127, 144, 162
Franklin, B. 182, 191
Frik, G. 106
Fystrom, D. 46
Gabor, D. 110
Gangadharan, D. 162
George, K. 127
Gould, M. 46
Green, D. 106
Gren, P. O. 106
Grofé, F. 202
Halsted, M. 130
Hampton, D. S. 121, 127, 180
Hansen, U. J. 127, 180
Hartenstein, W. 200, 201, 202
Hemony, F. and P. 128, 130, 140
Henderson, M. 78
Henze, H. W. 105
van Heuven, E. W. 128, 144
Holz, D. 53, 63
Hockstein, S. 76, 189, 195
Hoffman, B. 184
Holland, J. 106
Honegger, A. 105

Hopkin, B. 191, 202
Horner, A. 78
Houtsma, A. J. M. 10, 20, 25
Huygens, C. 128
Jiang, Z. 180
Kauffman, R. A. 78
Kalnins, A. 88
Kennedy, C. 127
Khachaturian, A. 74, 105
Kim Y. 178, 181
King, A. 78
Kirchgessner, M. 183
Kong, C. D. 178, 181
Kubik, G. 78
Kuttner, F. 198, 202
Kvistad, Garry v, vii, 63, 73
Kwon, J. 46
Lawson, J. R. 144
Lee, B. 181
Lee, J. M. 181
Legge, K. 106
Lehr, A. vii, 144, 145, 180, 202
Leissa, A. W. 88
Leung, K. K. 127
Levine, D. 446
Lieberman, R. 202
Lindsay, J. 46
Lundberg, L. 189
MacCallum, F. K. 59, 63
Mahler, G. 162
Maldonado, J. G. 127
Malm, W. P. 78
Malta, J.H. vii, 162
Mannette, E. 107, 123, 127
Mansell, E. R. 162
Marshall, B. 107
Mesmer, F. 183
Meyer, J. 58, 63
Mills, R. I. 20
Moore, J. 59, 63

Moussorgsky, M. 97
Murr, L. E. 123, 127
Musser, C. 54, 59
Nickerson, L. M. 106
Noonan, J. P. 46
Obata, J. 44, 46
O'Mahoney, T. 46
Orduña-Bustamante, F. 59, 63
Pappu, S. 127
Partch, H. 76, 189, 195
Peinkofer, K. 78
Perrier, A. 181
Perrin, R. 106, 144, 145, 162
Peterson, R. W. 162
Pichary, L. 126
Plomp, R. 10
Polo, M. 180
Posada, M. 127
Prak, A. 46
Price, P. 163, 181
Ramakrishna, B. S. 19, 20
Raman, C. V. 16, 20
Lord Rayleigh 7, 19, 67, 91, 128, 144, 162
Rehme, N. L. 191
Reissner, E. 88
Richardson, B. E. 180
Ritsma, R. J. 144
Roach, K. 126
Rockefeller, J. D. 129
Rohner, F. 122
Rose, C. 26, 46
Rossing, T. D. 10, 20, 25, 46, 51, 63, 78, 88, 105, 127, 144, 162, 163, 180, 181, 191
Rostoker, W. 181
Russell, D. A. 63, 163
Russell, D. P. 127
Saint-Saëns, C. 52
Sathoff, H. J. 162, 180
Satie, E. 202
Schad, C.-R. 106, 142, 145

Name Index

Schedin, S. 106
Schouten, J. F. 144
Schoenberg, A. 1, 105
Scott, J. F. M. 106
Seewan, 137, 144
Shepherd, R. B. 46, 78, 106
Simpson, A. 128, 144
Sinanan-Singh, K. A. 127
Sivian, L. J. 46
Slaymaker, F. H. 144
Smith, D. A. 191
Soedel, W. 181
Sondhi, M. M. 19, 20
Sons, W. 189
Sorin, E. 131
van Spiere, J. 130
Stevens, M. 196
Stiegler, E.-M. 191
Strauss, R. 202
Sykes, W. A. 20
Tannigel, F. 78
Taylor, H. W. 7, 19
Terhardt, E. 137, 144
Tesima, T. 44, 46
'T Hart, J. 144
Till, A. M. 202
Till, W. 199
Timoshenko, 67
Touzé, C. 106
Tracey, A. 77, 78
Trasher, A. R. 163
Trillo, E. A. 127
Tsai, J. 163, 180
Tubis, A. 20
Turner, J. 183
Tyzzer, F. 133, 144
Wagenaars, W.M. 10, 20, 25
Waller, M. D. 88
Warburton, G. B. 88, 162
Warlimont, H. 142, 145

White, S. D. 46
Wilbur, C. 106
Williams, A. 1007
Woodhouse, J. 106
Yum, Y. H. 178, 181
Zhao, H. 46, 180
Zheng D. 180
Zwislocki, J. J. 25

Subject Index

African drums 40
Alembas 71
Aluminum harp 194
Aluminum piano 197
Anklung 192
Armonica 182
Balafon 190
Bar, vibrations of 47
Bass drum 34
Bell lyra 65
Bell plates 104
Bell towers 132
Bells
  carillon 133
  church 133
  clappers 143
  handbells 146
  major-third 139
  scaling 139
  strike note 137
  temple 171
  tuning 136
  two-tone (Chinese) 164
  vibrational modes 133
  warble 142
Bi disk (pi disk) 198
Bodhrán 40
Bonang 1
Bongos 36
Boos (bamboo marimba) 195
Bronze drums 101
Carillons 128
Celesta 66
Ceramic instruments 200
Chladni patters 17
Chimes 67
  jade 198
  stone 198

Chinese opera gongs 100
Chinese two-tone bells 164
  bo 165
  interval between tones 168
  nao 165
  niuzhong 165
  yongzhong 165
  vibrational modes 166
Choirchimes 154
  vibrational modes 154
  sound spectra 157
Church bells 128
"Cloud chamber" bowls 189
Conga drums 35
Cow bells 161
Cristal 196
Crotals 160
Crotales 102
Cymbals 89
  nonlinear behavior 94
  sound 92
  vibrational modes 89
Crystallophone 191
Dipole 13
Djembé 40
End correction 51
Flexatone 105
Gamelan 71
Gamelan chimes 74
Gangsa 71
Gender 71
  gender panerus 72
  gender pemade 72
Glass bells 188
Glass harmonica 182
Glass harp 182
Glass orchestra 189
Glockenspiel 65
Gongs 98

Subject Index

Handbells 146
  bass 153
  scaling 152
  sound radiation 149
  timbre and tuning 150
  vibrational modes 147
Holographic interferometry 110
Indonesian drums 45
Japanese bells 178
Japanese drums 42
Jegogan 72
Kalimba 77
Kendang cilon 45
Kendang gending 45
Kenong 71
Ketuk 71
Korean bells 174
Kotodumi 42
Kyezee 103
Lamellaphones 77
Likembe 77
Lithophones 197
Loudness 22
Mallets 62
Marimbas 52
  orchestras 62
  resonators 60
  tuning the bars 58
Mark tree 196
Mbira 77
Membranes 6
Metallophones 65
Monopole 13
Mrdanga 15
Musical saw 104
O-daiko 42
Orchestra bells 64
Organ chimes 192
Pelog 71
Pentangles 69
Percussion ensembles 3
Pitch glide 26

Pitch of the missing fundamental 9
Quadrupole 13
Qing 158
Rototoms 37
Saron 71
Setinkane 78
Slendro 71
Slentem 71
Snare drum 28
  shell vibrations 29
  sound radiation 3
  snare action 33
Sound pressure level 23
Sound power level 23
Sound waves 22
Spoils of war 195
Steelpans 108
  construction and tuning 108
  double second 115
  metallurgy and heat treatment 122
  normal modes of vibration 109
  note shapes 121
  skirts 123
  sound spectra 117
  tenor (lead) 112
Stone chimes 198
Strings 5
Tabla 15
Talking drums 41
Tam-tams 96
Temple bells 171
  Burmese 180
  Chinese 171
  Japanese 178
  Korean 174
Thunder sheet 202
Timbales 36
Timpani 7
  kettle 10
  sound decay 12
Tom toms 27
Triangles 69

Tsuzumi (ko-tsuzumi, o-tsuzumi) 43
Tubaphones 74
Tubular bells 67
Tudumi 43
Turi-daiko 42
Typewriter 202
Verrophone 187
Vibraphone (vibes) 66
Vibrations
  of air columns 49
  of bars 48
  of membranes 6
  of plates 79
  of shells 83
  of strings 5
  of a wineglass 184
  torsional 49
Virtual pitch 9
Wind chimes 73
Xylophones 52
Zoomoozophone 76